Springer Proceedings in Mathematics & Statistics

Volume 24

For further volumes:
http://www.springer.com/series/10533

Springer Proceedings in Mathematics & Statistics

This book series features volumes composed of select contributions from workshops and conferences in all areas of current research in mathematics and statistics, including OR and optimization. In addition to an overall evaluation of the interest, scientific quality, and timeliness of each proposal at the hands of the publisher, individual contributions are all refereed to the high quality standards of leading journals in the field. Thus, this series provides the research community with well-edited, authoritative reports on developments in the most exciting areas of mathematical and statistical research today.

Bourama Toni • Keith Williamson • Nasser Ghariban
Dawit Haile • Zhifu Xie
Editors

Bridging Mathematics, Statistics, Engineering and Technology

Contributions from the Fall 2011 Seminar on Mathematical Sciences and Applications

 Springer

Editors
Bourama Toni
Department of Mathematics and Computer
    Sciences
Virginia State University
Petersburg, Virginia
USA

Nasser Ghariban
Department of Engineering
Virginia State University
Petersburg, Virginia
USA

Zhifu Xie
Department of Mathematics and Computer
    Science
Virginia State University
Petersburg, Virginia
USA

Keith Williamson
Department of Technology
Virginia State University
Petersburg, Virginia
USA

Dawit Haile
Department of Mathematics and Computer
Science
Virginia State University
Petersburg, Virginia
USA

ISSN 2194-1009   ISSN 2194-1017 (electronic)
ISBN 978-1-4899-9511-7   ISBN 978-1-4614-4559-3 (eBook)
DOI 10.1007/978-1-4614-4559-3
Springer New York Heidelberg Dordrecht London

Mathematics Subject Classification (2010): 97M10; 70F10; 51M04; 74R99; 97R99; 97R30; 82D80; 91A80; 93C95

Printed on acid-free paper

Springer is part of Springer Science+Business Media (www.springer.com)

Preface

This volume features selected and peer-reviewed contributions from the Fall 2011 Seminar Series on Mathematical Sciences and Applications at Virginia State University.

The weekly yearlong seminar is a dynamic forum for the best faculty to present and discuss current cutting-edge research at the interface between mathematics, science, engineering, and technology. The emphasis in this volume is on dynamical systems and the N-body problem in celestial mechanics, fractional calculus, almost and pseudo-almost periodicity, pseudo-almost automorphy, difference equations, calculus of variations and convexity, perfect polygons, and convex quadrics in geometry, as well as applications to materials sciences and pattern formation in microstructures, life sciences, computer science, bioinformatics and health, game theory, and economics. The participants, both faculty and students, are from several disciplines. The main objective is fostering student interest in STEAM-H (Science, Technology, Engineering, Agriculture, Mathematics and Health), stimulating graduate and undergraduate research and collaboration among researchers on a genuine interdisciplinary basis. Therefore, all articles are from contributors who are leading researchers in their field, and are carefully selective, self-contained, and pedagogically exposed.

The seminar takes place in an area that is socially, economically, intellectually very dynamic, and home to some of the most important research centers in the USA, including NASA Langley Research Center, manufacturing companies (Rolls-Royce, Canon, Chromalloy, Sandvik, Siemens, Sulzer Metco, NN Shipbuilding, Aerojet) and their academic consortium (CCAM), University of Virginia, Virginia Tech and Virginia State University, the Virginia Logistics Research Center, Virginia Nanotechnology Center, Aerospace Corporation, C3I Research and Development Center, Defense Advanced Research Projects Agency, Naval Surface Warfare Center, National Accelerator Facility, and the Homeland Security Institute. The program invites leading national and international researchers. The seminar is expected to become a national and international reference in STEAM-H education and research. To ensure thematic continuation, the proceedings are published by Springer, a renowned publisher with high standards for quality. The volume is organized as follows:

Chapter 1, by Dr. Zhifu Xie, is in celestial mechanics and presents current research on central configurations of the N-body problem, along with super central configurations, with some interesting applications to spacecraft orbital design.

Chapter 2, by Professor Gaston N'Guérékata, is in the rapidly expanding area of fractional differential equations, generalization of ordinary differential equations to arbitrary (non-integer) order capturing nonlocal relations in space and time. The author proves the existence of solutions to some Cauchy problems with nonlocal conditions.

Chapter 3, by Professor Toka Diagana, addresses nonautonomous systems of second-order differential equations. Using dichotomy tools the author obtains the existence of doubly weighted pseudo-almost automorphic solutions. An illustration of the non-resonance case is also presented.

Chapter 4, by Professor Stephen Schecter, applies mathematics to game theory and economics. The author revisits the biblical problem of how the Babylonian Talmud divides an estate among creditors when the debts total more than the estate using a game theoretical approach. He explains the Aumann–Maschler–Kaminski solutions and their relation to game theory (Robert Aumann is the 2005 Nobel prize winner in Economics.).

Chapter 5, by Professor Gaston N'Guérékata, is concerned with pseudo-almost periodicity and proves the nonuniqueness of the decomposition of a weighted pseudo-almost periodic function under certain conditions on the weight.

Chapter 6, by Dr. Paul Bezandry, studies almost periodic random sequences and their applications to the stochastic Beverton–Holt difference equation, in which the recruitment function and the survival rate vary randomly.

Chapter 7, by Dr Candace Kent, features an open problem in difference equations, in particular difference equations both piecewise-defined and having every solution either eventually periodic or unbounded. Four such cases and their common properties are studied. The chapter concludes with a very interesting query on the $3x + 1$ problem.

Chapter 8, by Dr Daniel Vasiliu, considers mathematical models for pattern formation in microstructures, applied to solid–solid phase transitions in materials science. Generalizations of rank-one convexity and quasi-convexity are developed and presented.

Chapter 9, by Dr. Kostadin Damevski, is devoted to the efficient programming of graphics processing units (GPU). The author develops a so-called refactoring tool, the ExtractKernel, to reduce the complexity of programming or porting codes to GPUs.

Chapter 10, by Dr. Weidong Mao, deals with bio-informatics. Data mining tools based on the so-called Association studies are used to efficiently find associations, patterns, and relationships on data collected about three different types of cancer patients. The study is based on data mining algorithms, and with the assistance of Oracle Data Miner Software, it describes the disparities on cancer survivability between African–American and White American populations

Chapter 12, by Professor Raymond Fletcher, features the so-called perfect hexagons and elementary triangles and their appearance in geometry. Drawing from abstract algebra and algebraic geometry, the author develops a theory of cubic curves with several esthetic and artful geometric representations. The many interesting results also lead to conjectures for future work.

The concluding chapter by Professor Valeriu Soltan investigates convex quadrics by extending the special class of ellipses and ellipsoid convex surfaces and classifies them in the n-dimensional real Euclidean space. Convex quadrics are also characterized using planes and hyperplanes quadric sections among all hypersurfaces, possibly unbounded.

This volume stimulates new advances in the fields of mathematics represented here, as well as in their applications in engineering, life and health sciences, game theory, and economics. The shared emphasis of these carefully selected and refereed contributed chapters is on important methods, research directions, and applications of analysis within and beyond mathematics. The seminar promotes mathematics, engineering, and technology education as well as interdisciplinary, industrial, and academic cooperation, through its Springer-published proceedings. The volume will serve as a source of inspiration for a broad spectrum of researchers and research students; the invited-only contributions are at the interface between modern mathematics and its applications in Science, Technology, Engineering, Agriculture, Mathematics, and Health, the so-called emerging STEAM-H interconnected disciplines.

We gratefully acknowledge the following supports: the Office of the Provost, Dr. Weldon Hill, Mr. Daniel Roberts, Ms. Yancey Dot, and Ms. Marie Singfield; the Office of the Dean, School of Engineering, Science and Technology, Professor Larry Brown, Ms. Victoria Perkins, Mrs. Bonnie Grant, and Mrs. Rudine Jenkins; the Department of Mathematics and Computer Science, its Chair Professor Kenneth Bernard and administrative assistants Ms. Caroline Price and Ms. Vickie Crowder; the Department of Education HBCU MS Program and Professor Pamela Leigh-Mack; the NSF/HBCU-UP, Professor Ali Ansari and Ms. Amber Dollete; the NIH/RIMI program and Dr. Omar Faison.

We would like to thank very much Ms. Melissa Watts, and also Dr. Giti Javidi, Dr. Tony Bryant, Mr. Leroy Lane, Mr. Daniel Huang, Ms. Eleanor Poarch-Wall, Mr. Daniel Fritz, Mr. Andrew Wynn, Ms. Owens Azzala, and Mr. Calvin Smith, whose tireless efforts have contributed to the smooth organization, presentations, recording, and attendance. We sincerely appreciate the promotional support by the Office of Students Activities and its Director Ms. Martin Menjiwe, and VSU Radio Station WVST 91.3 and Station Manager Ms. Jennifer Williamson and her assistants Ms. Melony Negron and Ms. Melissa Thornton.

Special thanks are extended to all the contributors, the faculty and student participants, in particular to Professor Oliver Hill, Professor Rana Singh, Professor Emeritus Walter Elias, Dr. Brian Sayre, Dr. Ehsan Sheybani, Professor Toka

Diagana, Dr. Ahmed Mohamed, and Professor Gaston N'Guérékata, great support-ers and/or frequent presenters at the seminar since its inception. We would like to express our sincere thanks to all the anonymous referees for their professionalism. They all made the seminar and its published thematic continuation a reality for the greater benefice of the community of science, engineering, technology, and mathematics.

Petersburg, VA, USA   Bourama Toni

Contents

Contributors

Paul H. Bezandry Department of Mathematics, Howard University, Washington, DC, USA

Kostadin Damevski Department of Mathematics and Computer Science, Virginia State University, Petersburg, VA, USA

Toka Diagana Department of Mathematics, Howard University, Washington, DC, USA

Raymond R. Fletcher III Department of Mathematics & Computer Science, Virginia State University, Petersburg, VA, USA

Candace M. Kent Department of Mathematics and Applied Mathematics, Virginia Commonwealth University, Richmond, VA, USA

Lisa Walls and Weidong Mao Department of Mathematics and Computer Science, Virginia State University, Petersburg, VA, USA

Gaston M. N'Guérékata Department of Mathematics, Morgan State University, Baltimore, MD, USA

Stephen Schecter Department of Mathematics, North Carolina State University, Raleigh, NC, USA

Valeriu Soltan Department of Mathematics, George Mason University, Fairfax, VA, USA

Daniel Vasiliu Department of Mathematics, Christopher Newport University, Newport News, VA, USA

Zhifu Xie Department of Mathematics and Computer Science, Virginia State University, Petersburg, VA, USA

Chapter 1
Central Configurations, Super Central Configurations, and Beyond in the n-Body Problem

Zhifu Xie

Abstract In this survey, we review our recent understandings on central configurations, super central configurations, and their applications. At the end, some challenging problems and further possible extensions are presented.

Keywords Central configurations • Super central configurations • n-body problems • Celestial Mechanics

AMS classification number: 70F10, 37J45, 37N05

1.1 n-Body Problem and Central Configurations

One of the oldest dreams of humankind is to understand the motion of celestial bodies such as the Sun, planets, and the Moon in the solar system. In 1686, Isaac Newton [17] described such motion as a solution of differential equations in his masterpiece "Philosophiae Naturalis Principia Mathematica." Most celestial bodies can be modeled as point masses. The Newtonian gravitational law asserts that the force acting on each body is directly proportional to the product of the masses, inversely proportional to the square of the distance between their centers, and acts along the straight line joining them. Then the motion of the particles can be stated as following:

Z. Xie (✉)
Department of Mathematics and Computer Science, Virginia State University,
Petersburg, VA 23806, USA
e-mail: zxie@vsu.edu

B. Toni et al. (eds.), *Bridging Mathematics, Statistics, Engineering and Technology,*
Springer Proceedings in Mathematics & Statistics 24, DOI 10.1007/978-1-4614-4559-3_1,
© Springer Science+Business Media New York 2012

$$m_k \ddot{q}_k = \frac{\partial U}{\partial q_k} = \sum_{j=1, j \neq k}^{n} \frac{m_k m_j (q_j - q_k)}{|q_j - q_k|^3} \qquad 1 \leq k \leq n, \qquad (1.1)$$

where $m_k \in \mathbf{R}^+$ is the mass of k-th body with position $q_k \in \mathbf{R}^3 (k = 1, \cdots, n)$, and U is the Newtonian potential function

$$U = \sum_{1 \leq k < j \leq n} \frac{m_k m_j}{|q_k - q_j|}. \qquad (1.2)$$

We will use $q \in (\mathbf{R}^3)^n$ and $m \in (\mathbf{R}^+)^n$ to denote the position and mass vectors $(q_1, \cdots, q_n)$ and $(m_1, \cdots, m_n)$, respectively. Let $C = m_1 q_1 + \cdots + m_n q_n$, $M = m_1 + \cdots + m_n$, $c = C/M$ be the first moment, total mass, and center of mass of the bodies, respectively.

The differential equations (1.1) of the two-body (n = 2) problem are easy to solve. Its complete solution was first given by the Swiss mathematician John Bernoulli in 1710. One can prove that the path followed by one particle with respect to the other always lies along a conic section. This implies that the orbit described in physical space may be a circle, an ellipse, a parabola, a branch of hyperbola, or a straight line. Unfortunately the two-body problem has proved to be the only easy one among all the n-body problems; in spite of an enormous expenditure of effort, for n larger than two, even for $n = 3$, no other case has been solved completely.

The major breakthrough was achieved by Henri Poincaré in late nineteeth century. His main result shows that it is not possible to find a general solution of the systems and that the behavior of general solutions can be chaotic. This is the starting point of qualitative theory of dynamical systems which has thrived in the twentieth century. Even so, the n-body problem remains one of the favorite problems of leading mathematicians. As pointed out by Jürgen Moser [14] in his 1998 ICM hour-long talk, the n-body problem is the driving force in the development of dynamical systems and other areas. The focus has been switched from seeking general solutions to finding special periodic solutions and investigating initial configurations with special behavior. A central configuration plays the essential role in understanding the global structure of solutions of n-body problem of celestial mechanics. A central configuration is an arrangement of the initial positions of masses that leads to special families of solutions of the n-body problem. The initial positions must satisfy the following nonlinear system of algebraic equations:

$$\sum_{j=1, j \neq k}^{n} \frac{m_j (q_j - q_k)}{|q_j - q_k|^3} = -\lambda (q_k - c), \qquad 1 \leq k \leq n, \qquad (1.3)$$

for a constant λ. By the homogeneity of $U(q)$ of degree -1, $\lambda = U/2I > 0$, where I is the moment of inertial of the system, i.e., $I = \frac{1}{2} \sum_{i=1}^{n} m_i |q_i|^2$. The collision set is defined by

$$\triangle = \bigcup \{ q = (q_1, q_2, \cdots, q_n) \in (\mathbf{R}^3)^n | q_i = q_j \text{ for some } i \neq j \}. \qquad (1.4)$$

To avoid singularities q is restricted in $V(n)$:

$$V(n) = \{q = (q_1, q_2, \cdots, q_n) \in (\mathbf{R}^3)^n\} \backslash \triangle. \qquad (1.5)$$

Definition 1.1.1 (Central Configuration). A configuration $q \in V(n)$ is a *central configuration* (CC for short) for a given mass vector $m = (m_1, m_2, \cdots, m_n) \in (\mathbf{R}^+)^n$ if q is a solution of the system (1.3) for some constant $\lambda \in \mathbf{R}$.

The study of central configurations plays a "central" role in the understanding of the complexity of the n-body problem and it is a subject that develops in many directions. The problem of finiteness of the number of central configurations has been listed as a challenging problem for 21st century's mathematicians (see S.Smale [24]). The relevance of central configurations is remarked in multiple works in the literature. We refer specially to the beautiful presentation by Sarri [20, 21], Meyer-Hall [12], Meyer-Hall-Offin [13], and Moeckel [16] where the fundamental references are found.

Central configurations lead to the only known cases where the differential equations of the n-body problem are integrable. This was already known in Newton's era. Any planar central configuration gives rise to a special one-parameter family of periodic solution where each body rotates around the center of mass on its own ellipse. If all the ellipses are circles, the solution is known as a relative equilibrium because it is a fixed point in a rotating coordinate system. Some of the earliest solutions to the three-body problem were of this type. For example, Euler [5] in 1767 proved that there exists exactly one central configuration for each ordering of the three masses on a line. Lagrange [6] in 1772 discovered the equilateral triangle central configurations. When three bodies with any choice of masses are placed at the vertices of an equilateral triangle, it gives rise to a family of solutions where each body is traveling along a particular Kepler orbit.

Indeed twentieth century astronomers later discovered two groups of asteroids (now called the Trojans in 1906 and the Eureka in 1990) which form an equilateral triangle with the Sun and Jupiter. Central configurations have been applied in spacecraft mission design. When a spacecraft is placed at a location where the Earth, the Moon, and the spacecraft form a central configuration, all the forces among them are exactly balanced and the spacecraft will stay there forever in the rotating system. These locations are called Euler–Lagrange points or libration points. This is important in applications, since a spacecraft can be theoretically parked at one of the Euler–Lagrange points without any fuel for as long as wanted. Linear stability analysis on the equilibrium points shows that there exists a weak stability boundary (WSB), where the gravitational forces from the Earth and the Moon are nearly balanced in relation to a moving spacecraft as it travels around the Moon. It provides a new useful mission design with low energy transfer (ballistic capture transfer or WSB transfer) which is indeed used in the rescue of a spacecraft probe to the Moon [4].

In the concept of low energy interplanetary transfers [7,8], the role of the libration points is the key. However the mathematical theory behind the idea is very simple. It is based on the understanding of the nonlinear dynamics of these orbits by the invariant manifold theory. The solar system is modeled by a series of circular restricted three body problems. Each of them gives rise to some libration points. The libration points of all the planets and their moons generate some stable and unstable manifolds which are called tunnels. Then our solar system is interconnected by a vast system of tunnels winding around the Sun. Our spacecraft can fly through the tunnels with fewer energy consumption.

1.2 Super Central Configurations

A new phenomenon of central configurations was recently discovered in the collinear three-body and four-body problem. There exists a configuration that is a central configuration for at least two different arrangements of a given mass vector. Given a configuration $q = (q_1, q_2, \cdots, q_n) \in V(n)$, denote $S(q)$ the admissible set of masses by

$$S(q) = \{m = (m_1, m_2, \cdots, m_n) | m_i \in \mathbf{R}^+, q \text{ is a central configuration for } m\}. \quad (1.6)$$

For a given $m \in S(q)$, let $S_m(q)$ be the permutational admissible set about m, denoted by

$$S_m(q) = \{m' \in S(q) | m' \neq m \text{ and } m' \text{ is a permutation of } m\}. \quad (1.7)$$

The requirements that $m' \neq m$ and m' is a permutation of m in $S_m(q)$ are necessary to exclude some trivial cases. For example, if q is a central configuration for $m = (m_1, m_2, m_3, \cdots, m_n)$ with $m_1 = m_2$, then q is also a central configuration $m' = (m_2, m_1, m_3, \cdots, m_n)$ but $m' \notin S_m(q)$. Let $P(n)$ be the set of permutations of $\{1, 2, \cdots, n\}$. The set $S_m(q)$ is a finite subset of $\{m(\tau) | \tau \in P(n)\}$ and has at most $n! - 1$ elements in $S_m(q)$.

Definition 1.2.2. Configuration q is called a *super central configuration* (SCC for short) if there exists positive mass m such that $S_m(q)$ is nonempty.

The existence of super central configurations is a special inverse problem of central configurations. The first example of super central configuration is the equilateral triangle configuration in the planar three-body problem. If q is the equilateral triangle configuration and $m = (m_1, m_2, m_3)$, then q is also a central configuration for each permutation of m. Therefore, for three distinct masses, the set $S_m(q)$, which has five elements, consists of all the permutations of (m_1, m_2, m_3). However, there is no super central configuration in planar four-body problem.

The existence and classification of super central configurations has been studied in the collinear three-body problem [25] and in the collinear four-body problem [26]. The detailed classifications of the nonempty set $S_m(q)$ and the exact configurations of super central configurations are also established. In the collinear three-body problem, to find the exact configuration of super central configurations, the general configuration up to translation and scaling is chosen at the x-axis $q = ((0,0,0), (1,0,0), (1+r,0,0))$. Albouy–Moeckel [3] found that the center of mass does not depend on the choice of the masses which make the configuration central and this property was also given by Christian Marchal in his book [11]. By using the remarkable property, that $^{\#}S_m(q) \leq 1$ is proven analytically for any configuration and any mass $m = (m_1, m_2, m_3)$. Moreover q is a super central configuration when $\underline{r} < r < 1/\underline{r}$ and $r \neq 1$, where $\underline{r}$ is a unique positive zero of a polynomial with degree 6. The set $S_m(q)$ can only be either $\{(m_3, m_1, m_2)\}$ or $\{(m_2, m_3, m_1)\}$. Such results (see [27]) have been extended by an undergraduate research group at the Virginia State University to the case of inverse integer power law.

In the paper [28] entitled by "The Golden Ratio and Super Central Configurations of the n-body Problem," it is further discovered that golden ratio $\phi = (\sqrt{5}+1)/2$ has surprising connections with super central configurations of the N-body problem. It is amazing that this mathematical beauty is hidden in the action of celestial particles. Let $r = (|q_3 - q_2|)/(|q_2 - q_1|)$ be the ratio of distances in the collinear three-body problem with the ordered positions q_1, q_2, q_3 on a line. Only if r is greater than $1/\phi$ and less than ϕ, the configuration could be a super central configuration.

In the collinear four-body problem, the general configuration is chosen at the x-axis $q = ((-s - 1, 0, 0), (-1, 0, 0), (1, 0, 0), (t + 1, 0, 0))$. Ouyang–Xie [18] gave explicit expressions of the masses which make the configuration central. The center of mass does depend on the choice of the masses. Surprisingly, it was proved in [26] that the constant λ in (1.3) does not depend on the order of the bodies for any elements in $S_m(q)$. By using this property and the linearity of the center of mass, that $^{\#}S_m(q) \leq 1$ is proven analytically for any configuration and any mass $m = (m_1, m_2, m_3, m_4)$. To find the exact super central configurations and the corresponding masses, a polynomial equation in two position variables s, t is derived. The polynomial equation that consists of 291 terms with degree 29 is quite complicated. Amazingly, it is proved just using Descartes' Rule that the polynomial equation gives rise to a unique curve in the st-plane which must be satisfied by any super central configuration.

Based on the results in the collinear three-body and four-body problems, we pose the following questions. Some of these questions have been reported in the 2011 AMS Spring Eastern Sectional Meeting at the College of Holly Cross.

Question One: How large can the set $S_m(q)$ be?

Question Two: Will a super central configuration give a special type of "perverse solutions" with some special type of dynamical behaviors?

Conjecture: $^{\#}S_m(q)$ is either zero or one for the collinear n-body problems.

1.3   Super Central Configurations and Number of Central Configurations

For any given mass vector $m \in (\mathbf{R}^+)^n$, the sets $L_G(n, m)$, $L_P(n, m)$, and $L_M(n, m)$ denote the set of all geometric equivalence, permutation equivalence, and mass equivalence classes of n-body collinear central configurations, respectively.

Most papers and books study the central configurations under permutation equivalence. Generally speaking, permutation of bodies makes difference in permutation equivalence and geometric equivalence refers to the equivalence of geometric shapes. More details on the definitions of the equivalence can be found in the paper [19].

As the results of [1, 2] for the four equal masses case, there are exactly four planar central configurations under geometric equivalence, i.e., the square, a special isosceles triangle with one body on its axis of the symmetry, an equilateral triangle with one body at its center, and a collinear central configuration. However, there are 50 central configurations for the four equal masses case under permutation equivalence. This gives a good example on which the number of central configurations is different under the different equivalence.

Directly from the definitions, we can deduce that $^\#L_G(n,m) \leq^\# L_M(n,m) \leq^\# L_P(n,m)$. Under the definition of permutation equivalence of central configurations, collinear central configurations are one of a few families of central configurations with given positive masses which are somewhat understood. Moulton [15] proved that there is a unique position that causes a central configuration for each way the particles can be ordered along a line in 1910 and Smale [22] reconfirmed the result by a different variational approach in 1970. Therefore, $^\#L_P(n,m) = n!/2$.

The decreasing phenomenon of the number of central configurations has been observed in a long history. The decreasing phenomenon in $^\#L_M(3,m)$ was studied by Wintner [24] in 1941. For example $^\#L_P(3,m) = 3$ but $^\#L_M(3,m) = 2$ if two of m_1, m_2, m_3 are equal but not the third. Long–Sun [9, 10] established the counting number $^\#L_M(n,m)$ in the collinear n-body problem.

But $^\#L_G(n,m)$ is only known in the collinear three-body problem and in the collinear four-body problem. Long–Sun [9, 10] first addressed the problems and they gave results on the enumerations of central configurations under each equivalence, especially, in the sense of geometric equivalence they found a singular algebraic hypersurface in the mass space which decreases the number of central configurations in the three-body problem. It was proved $^\#L_P(3,m) =^\# L_M(3,m) = 3$ but $^\#L_G(3,m) = 2$ if m is in the singular algebraic hypersurface. This paper [29] reinvestigated the case of collinear central configurations of the three-body problem. The paper provided a direct parametric expression for the singular algebraic hypersurface in the mass space and a different proof involving the information of super central configurations. Ouyang and Xie [19] found the exact number of central configurations of the collinear four-body problem under geometric equivalence. The expression of the singular algebraic hypersurface in the mass space only depends on two parameters. If m is in the singular algebraic hypersurface, then $^\#L_P(4,m) =^\# L_M(4,m) = 12$ but $^\#L_G(4,m) = 11$. If m is not in the singular algebraic hypersurface, $^\#L_G(4,m) =^\# L_M(4,m)$.

Question Three: What is the Relationship Between the Super Central Configurations and the Number of Central Configurations under Geometric Equivalence?

If the masses are distinct, the decreasing phenomenon is closely related to the existence of super central configurations, i.e., for some distinct masses m, $^\#L_G(n,m)$

is strictly less than $^{\#}L_M(n,m)$. The difficulties of the proofs in [19, 29] are to check the equivalence of central configurations for many different arrangements of the bodies. For example, more than 90 cases have been checked even after some subtle simplifications with some remarkable properties in the collinear four-body problem. The possible cases to be checked will increase tremendously as n increases due to the fact that the number of permutations of n-body is $n!$. Previous results with $n = 3, 4$ show that the decreased number of central configuration under geometric equivalence is solely due to the existence of super central configurations if masses are distinct. It is possible to find the number of central configurations under geometric equivalence by excluding the number of super central configurations. Establishing the relationship between super central configurations and the number of central configurations is an important step towards understanding the number decreasing phenomenon and it will provide an easier way to count the number of central configurations.

Acknowledgments The author would like to thank Professor Bourama Toni for his encouragements and suggestions regarding this work.

References

1. Albouy, A.: Symetrie des configurations centrales de quatre corps. (French) [Symmetry of central configurations of four bodies] C. R. Acad. Sci. Paris Soc. I Math. **320**(2), 217–220 (1995)
2. Albouy, A.: The symmetric central configurations of four equal masses. Hamiltonian dynamics and celestial mechanics (Seattle, WA, 1995), Contemporary Mathematics, American Mathematical Society, Providence, RI, **198**, pp.131–135 (1996)
3. Albouy, A., Moeckel, R.: The inverse problem for collinear central configuration, Celestial. Mech. Dyn. Astron. **77**, 77–91 (2000)
4. Belbruno, E.: Fly me to the Moon. Princeton University Press (2007)
5. Euler, L. : De motu rectilineo trium corporum se mutuo attahentium. Novi Comm. Acad. Sci. Imp. Petrop. **11**, 144–151 (1767)
6. Lagrange, J.: Essai sur le problème des trois corps. Euvres, vol. 6, pp. 272292. Gauthier-Villars, Paris (1772)
7. Lo, M., Ross, S.: Surfing the Solar System: Invariant manifolds and the dynamics of the solar System, JPL IOM 312/97 (1997)
8. Lo, M., Ross, S.: The Lunar L1 Gateway: Portal to the Stars and Beyond, AIAA Space 2001 Conference, Albuquerque, NM, 28–30 August (2001)
9. Long, Y., Sun, S.: Collinear central configurations and singular surfaces in the mass space, Arch. Rational Mech. Anal. **173**, 151–167 (2004)
10. Long, Y., Sun, S.: Collinear central configurations in celestial mechanics. In: Brezis,H., Chang, K.C. Topological Methods, Variational Methods and Their Applications: ICM 2002 Satellite Conference on Nonlinear Functional Analysis, Taiyuan, Shan Xi, China, P.R. August 14–18, p. 159–165 (2002)
11. Marchal, C. The Three-Body Problem, p. 44. Elsevier, Amsterdam (1990)
12. Meyer, K., Hall, G.R.: Introduction to Hamiltonian Dynamical System and the N-Body Problem, Applied Mathematical Sciences, Vol. 90. Springer (1992)

13. Meyer, K., Hall, G.R., Offin, D.: Introduction to hamiltonian dynamical system and the N-Body problem. In: Applied Mathematical Sciences, vol. 90, 2nd edn. Springer (2009)
14. Moser, J.: Dynamical system-past and present. Regular Chaotic Dyn. 13(6), 499–513 (2008)
15. Moulton, F.R.: The straight line solutions of the problem of N bodies. Ann. Math., II. Ser. 12, 1–17 (1910)
16. Moeckel, R.: On central configurations. Math. Zeit. 205 499–517 (1990)
17. Newton, I.: Philosophiae Naturalis Principia Mathematica, S. Pepys. Royal Society Press, London (1686)
18. Ouyang, T., Xie, Z.: Collinear central confiugration in four-body problem. Cele. Mech. Dyna. Astr. 93, 147–166 (2005)
19. Ouyang, T., Xie, Z.: Number of central configurations and singular surfaces in the mass space in the collinear four-body problem. Trans. Am. Math. Soc. 364, 2909–2932 (2012)
20. Saari, D.: On the role and the properties of n body central configurations. Clestial Mech. 21, 9–20 (1980)
21. Saari, D.: Collisions, Rings, and Other Newtonian N-body Prolbems. American Mathematical Society, Providence (2005)
22. Smale, S.: Topology and mechanics.II. The planar n-body problem. Invent. Math. 11, 45–64 (1970)
23. Smale, S.: Mathematical problems for the next century. Math Intel. 20(2) (1998)
24. Wintner, A.: The analytical foundations of celestial mechanics. Princeton Math. Series 5, 215. Princeton Univ. Press, Princeton (1941)
25. Xie, Z.: Super central configurations of the n-body Problem. J. Math. Phys. 51, 042902 (2010)
26. Xie, Z.: Inverse problem of central configurations and singular curve in the collinear 4-Body Problem. Cele. Mech. Dyna. Astr. 107, 353–376 (2010)
27. Xie, Z., Hodge, K., Westbrook, M., Henderson K.: Super central configurations of the three-body problem under the inverse integer Power Law. J. Math. Phys. 52, 092901 (2011)
28. Xie, Z.: The Golden ratio and super central configurations of the n-body problem. J. of Diff. Equa. 251, 58–72 (2011)
29. Xie, Z.: Central configurations of collinear three-body problem and singular surfaces in the mass space. Phys. Lett. 375, 3392–3398 (2011)

Chapter 2
A Note on Fractional Calculus and Some Applications

Gaston M. N'Guérékata

Abstract In this short note, we show by simple examples that differential equations of fractional orders generalize the ones of integer orders. We present a variation of constants formula which we obtained recently with C. Lizama and use it to prove the existence of solutions to some Cauchy problem with nonlocal conditions. This latter generalizes some of our recent results.

Keywords Derivative of the fractional order • Mild solution

Mathematics Subject Classification (1991): Primary: 47D06; Secondary: 34G10, 45M05

2.1 Introduction

Fractional calculus is a three-century-old topic. It goes back to Leibniz. First, note that the notation $\frac{d^n y}{dt^n}$, $n = 1, 2, \ldots$ of the derivatives of a function y is due to Leibniz. In a 1965 correspondence to Leibniz, L'Hospital wrote: "what if $n = \frac{1}{2}$?." In his response, Leibniz said, "this is an apparent paradox from which, one day, useful consequences will be drawn."

Early contributors include Euler (1730), Lagrange (1772), Fourier (1822), Liouville (1832), Riemann (1847), Leitnikov (1868), Laurent (1884), Krug (1890), and Weyl (1917).

G.M. N'Guérékata (✉)
Department of Mathematics, Morgan State University, Baltimore, MD 21251, USA

Laboratoire CEREGMIA, Université des Antilles et de la Guyane,
97159 Point-à-Pitre, Guadeloupe (FWI)
e-mail: Gaston.N'Guerekata@morgan.edu

B. Toni et al. (eds.), *Bridging Mathematics, Statistics, Engineering and Technology*,
Springer Proceedings in Mathematics & Statistics 24, DOI 10.1007/978-1-4614-4559-3_2,
© Springer Science+Business Media New York 2012

Over the last decade, there has been a resurgence of fractional differential equations and their applications to science (cf. for instance [1–14] and references therein).

Indeed, fractional calculus is applied in almost all areas of science. One might cite for instance Mechanics (theory of viscoelasticity and viscoplasticity), (Bio) Chemistry (modeling of polymers and proteins), Electrical Engineering (transmission of ultrasound waves), Medicine (modeling of human tissue under mechanical loads), Mathematical Psychology (modeling of behavior of human beings), Control Theory (implementation of fractional order controllers), etc...Fractional differential equations are more appropriate and more efficient in the modeling of memory-dependent phenomena and the modeling in complex media, such as porous ones.

The paper is organized as follows. In Sect. 2.2, we present some elementary properties of the fractional derivatives in the sense of Caputo and in the sense of Riemann–Liouville. In Sect. 2.3, we solve some simple fractional differential equations using the Caputo and the Riemann–Liouville derivatives and compare the solutions to the one of a differential equation of order one. In Sect. 2.4, we present a variation of constants formula due to Lizama and N'Guérékata [7]. Finally in Sect. 2.5, we study the existence of solutions to a Cauchy problem with nonlocal conditions, using classical fixed-point theorems. These results generalize our recent ones in [13].

2.2 Fractional Derivatives

In this section, we will present the Caputo fractional derivative and the Riemann–Liouville fractional derivative, two concepts of fractional derivatives among the most used in the literature and recall some of their properties.

Definition 2.2.1. The fractional integral of order $\alpha > 0$ of a function f is defined as

$$J_t^\alpha f(t) := \frac{1}{\Gamma(\alpha)} \int_0^t (t-s)^{\alpha-1} f(s) \, ds$$

provided the right-hand side is pointwise defined on $[0, \infty)$.

Let $g_\alpha(t) := \frac{t^{\alpha-1}}{\Gamma(\alpha)}$ if $t > 0$ and $:= 0$ if $t \leq 0$.

Then

$$J_t^\alpha f(t) = (g_\alpha \star f)(t)$$

where

$$J_t^0 f(t) = f(t).$$

Here $(\cdot \star \cdot)$ denotes the convolution operator.

Let X be a Banach space with norm $\|\cdot\|$.

Definition 2.2.2. Let $f \in C^m(R^+, X)$. If $m - 1 < \alpha < m$ where $m \in \mathbb{N}$, the Riemann–Liouville derivative of f of order α is

$$D_{RL}^\alpha f(t) := \frac{d^m}{dt^m} \int_0^t \frac{(t-s)^{m-\alpha-1}}{\Gamma(m-\alpha)} f(s)ds = \frac{d^m}{dt^m} \int_0^t g_{m-\alpha}(t-s)f(s)ds$$

Definition 2.2.3. The Caputo derivative of f of order $\alpha \in (m-1, m)$ is

$$D_c^\alpha f(t) := \int_0^t \frac{(t-s)^{m-\alpha-1}}{\Gamma(m-\alpha)} f^{(m)}(s)ds$$

$$= \int_0^t g_{m-\alpha}(t-s)f^{(m)}(s)ds$$

$$= J_t^{m-\alpha} f^{(m)}(t).$$

Theorem 2.2.4. If $f = c$, a constant, then $D_c^\alpha f = 0$, but $D_{RL}^\alpha f \neq 0$.

Proof. If $f = c$, then we have $D_c^\alpha f = \int_0^t g_{m-\alpha}(t-s) \frac{d^m}{ds^m}(c)ds = 0$.
But

$$D_{RL}^\alpha f = \frac{d^m}{dt^m} \int_0^t g_{m-\alpha}(t-s)c\,ds = \frac{c}{\Gamma(m-\alpha)} \frac{d^m}{dt^m} \left(\frac{t^{m-\alpha}}{m-\alpha} \right)$$

$$= \frac{c(m-\alpha-1)(m-\alpha-2)\ldots(-\alpha+1)t^{-\alpha}}{\Gamma(m-\alpha)} \neq 0.$$

$\square$

The following provides a relation between the Caputo derivative and the Riemann–Liouville derivative (cf. for instance [14]).

Theorem 2.2.5. $D_c^\alpha h(t) = D_{RL}^\alpha h(t) - \frac{t^{-\alpha}}{\Gamma(1-\alpha)} h(0^+)$, $0 < \alpha < 1$.

Theorem 2.2.6. Let $\alpha > 0$. If $D_c^\alpha h(t) = 0$, then

$$h(t) = c_0 + c_1 t + c_2 t^2 + \cdots + c_n t^{n-1}$$

where c_i are reals and $n = [\alpha] + 1$.

2.3   Differential Equations

For $z \in \mathbb{C}$, $\alpha > 0$, $\beta > 0$, let

$$\mathcal{E}_{\alpha,\beta}(z) := \sum_{k=0}^\infty \frac{z^k}{\Gamma(\alpha k + \beta)}$$

be the two-parameter Mittag-Lefler function and set

$$\mathcal{E}_{\alpha,1}(z) := \mathcal{E}_\alpha(z).$$

Then one obtains the following:

- $\mathcal{E}_1(z) = e^z$
- $\mathcal{E}_2(z^2) = \cosh(z)$
- $\mathcal{E}_2(-z^2) = \cos(z)$

From the above, it is clear that the two-parameter Mittag-Lefler function is a generalization of the exponential function. It is an essential tool in the study of fractional differential equations.

Consider for $\lambda > 0$ the ordinary differential equation

$$y'(t) + \lambda y(t) = f(t)$$
$$y(0) = y_0 \tag{2.1}$$

The solution is given by

$$y(t) = \mathcal{E}_1(-\lambda t)y_0 + \int_0^t \mathcal{E}_1(-\lambda(t-s))f(s)\mathrm{d}s. \tag{2.2}$$

Now consider for $0 < \alpha < 1$ and $\lambda > 0$, the fractional differential equation

$$D_c^\alpha y(t) + \lambda y(t) = f(t)$$
$$y(0) = y_0 \tag{2.3}$$

Let $\hat{y}(s)$ and $\hat{f}(s)$ be the Laplace transforms of y and f, respectively. Taking the Laplace transform of Eq. (2.3), we get

$$s^\alpha \hat{y}(s) - s^{\alpha-1}y_0 + \lambda \hat{y}(s) = \hat{f}(s)$$

Thus

$$\hat{y}(s) = \frac{s^{\alpha-1}}{s^\alpha + \lambda}y_0 + \frac{\hat{f}(s)}{s^\alpha + \lambda}$$

Then using the inverse Laplace transform

$$\frac{t^{-\alpha}}{\Gamma(1-\alpha)} \to s^\alpha$$

$$t^{\alpha-1}\mathcal{E}_\alpha(-\lambda t^\alpha) \to \frac{1}{s^\alpha + \lambda}$$

gives the following solution to Eq. (2.3):

$$y(t) = \mathcal{E}_\alpha(-\lambda t^\alpha)y_0 + \int_0^t (t-s)^{\alpha-1}\mathcal{E}_\alpha(-\lambda(t-s)^\alpha)f(s)ds. \tag{2.4}$$

Consider now the fractional differential equation

$$D_{RL}^\alpha y(t) + \lambda y(t) = f(t)$$
$$(g_{1-\alpha} \star y)(0) = y_0 \tag{2.5}$$

Remark 2.3.1. Note here that the initial condition corresponds to $D_c^\alpha y(0) = 0$ in view of Theorem 2.2.5.

Now, taking the Laplace transform of Eq. (2.5) yields

$$s^\alpha \hat{y}(s) - y_0 + \lambda \hat{y}(s) = \hat{f}(s).$$

Thus

$$\hat{y}(s) = \frac{1}{s^\alpha + \lambda}y_0 + \frac{\hat{f}(s)}{s^\alpha + \lambda}$$

Taking the inverse Laplace transform gives the solution to Eq. (2.5)

$$y(t) = t^{\alpha-1}\mathcal{E}_\alpha(-\lambda t^\alpha)y_0 + \int_0^t (t-s)^{\alpha-1}\mathcal{E}_\alpha(-\lambda(t-s)^\alpha)f(s)ds \tag{2.6}$$

Remark 2.3.2. If we let $\alpha = 1$ in (2.4) and (2.6), we obtain (2.2). This proves that both Eqs. (2.3) and (2.5) generalize Eq. (2.1).

2.4   A Variation of Constants Formula

In this section, we present a variation of constants formula under minimal conditions. We assume that $A : D(A) \subset X \to X$ is a closed linear operator and $f, g : \mathbb{R} \times X \to X$ are continuous functions. Consider for $0 < \alpha < 1$ the fractional differential equation

$$D_c^\alpha(u(t) + g(t, u(t))) = Au(t) + f(t, u(t)), \quad u(0) = u_0 \tag{2.7}$$

and the integral equation

$$u(t) = u_0 + g(0, u_0) - g(t, u(t)) + \int_0^t g_\alpha(t-s)f(s, u(s))ds$$
$$+ A\int_0^t g_\alpha(t-s)u(s)ds, \tag{2.8}$$

or

$$u(t) + g(t, u(t)) = u_0 + g(0, u(0)) + J_t^\alpha f(t, u(t)) + A J_t^\alpha u(t).$$

Theorem 2.4.1. *Equations (2.7) and (2.8) are equivalent.*

Proof. First we have

$$D_c^\alpha f(t) = J_t^{m-\alpha} D_c^m f(t), \quad m = \lceil \alpha \rceil$$

where $J_t^\alpha = \int_0^t g_\alpha(t-s) f(s) \mathrm{d}s = (g_\alpha \star f)(t)$. The following are well known.

(1) $D_c^\alpha J_t^\alpha f = f$
(2) $J_t^\alpha D_c^\alpha f = f(t) - f(0)$

Now suppose that (1) holds. Apply J_t^α to both sides. We get

$$u(t) + g(t, u(t)) - (u(0) + g(0, u(0))) = A J_t^\alpha u(t) + J_t^\alpha f(t, u(t))$$

or

$$u(t) + g(t, u(t)) - (u(0) + g(0, u(0))) = A(g_\alpha \star u)(t) + (g_\alpha \star f(t, u(t)))$$

which is (2).

Conversely suppose (2) holds. Apply D_c^α to both sides of (2). We get

$$D_c^\alpha(u(t) + g(t, u(t))) = D_c^\alpha(u_0 + g(0, u(0))) + D_c^\alpha J_t^\alpha f(t, u(t)) + D_c^\alpha A J_t^\alpha u(t)$$

or

$$D_c^\alpha(u(t) + g(t, u(t))) = f(t, u(t)) + Au(t)$$

since $D_c^\alpha(\text{constant}) = 0$ and $D_c^\alpha J_t^\alpha h = h$. $\qquad\square$

2.5   A Cauchy Problem with Nonlocal Conditions

Let $X = \mathbb{R}^n, I = [0, T], \mathcal{C} := C(I, X)$ the Banach space of continuous functions : $I \to X$ endowed with the topology of uniform convergence $\| \cdot \|_{\mathcal{C}}$. Consider the Cauchy problem with nonlocal conditions

$$D_c^\alpha u(t) = f(t, u(t)), \ t \in (0, T]$$

$$u(0) + g(u) = u_0. \tag{2.9}$$

Here $0 < \alpha < 1$. We make the assumptions

- **H1** $f(t, u)$ is of Carathéodory, i.e., for any $u \in X$, $f(t, u)$ is strongly measurable with respect to $t \in I$, and for any $t \in I$, $f(t, u)$ is continuous with respect to $u \in X$.
- **H2** $\|f(t, x) - f(t, y)\| \le L\|x - y\|, \ \forall x, y \in X, \ \forall t \in I$

- **H3** $g : C \to X$ is continuous; moreover, there exists $b > 0$ such that

$$\|g(u) - g(v)\| \leq b\|u - v\|_C, \quad \forall u, v \in C.$$

We have the following results with slight generalizations of Theorems 2.1 and 2.2 [13], because of **H1**.

Theorem 2.5.1. *Under assumptions* **H1–H3**, *(A) has a unique solution provided* $b < \frac{1}{2}$ *and* $L < \frac{\Gamma(\alpha+1)}{2T^\alpha}$.

Proof. According to Theorem 2.4.1, Eq. (2.5) is equivalent to the integral equation

$$u(t) = u_0 - g(u) + \frac{1}{\Gamma(\alpha)} \int_0^t (t - s)^{\alpha-1} f(s, u(s)) \mathrm{d}s.$$

We proceed as in [13]. We define the operator $\mathcal{F} : C \to C$ by

$$(\mathcal{F}u)(t) := u_0 - g(u) + \frac{1}{\Gamma(\alpha)} \int_0^t (t - s)^{\alpha-1} f(s, u(s)) \mathrm{d}s.$$

and prove that $\mathcal{F}$ has a unique fixed point by the Banach contraction principle.   □

Theorem 2.5.2. *Suppose that* **H1–H3** *are satisfied with* $b < 1$. *Suppose also that*

$$\|f(t, x)\| \leq \mu(t), \ \forall (t, x) \in I \times X$$

where $\mu \in L^1(I, \mathbb{R}^+)$. *Then Eq. (2.5) has at least one solution on I.*

Proof. As in [13] Theorem 2.2, we use Krasnoselkii's theorem to achieve the conclusion.   □

References

1. Abbas, S., Benchohra, M., N'Guérékata, G.M.: Topics in Fractional Differential Equations. Springer, New York (2012)
2. Bagley, R.L., Torvik, P.J.: On the appearance of the fractional derivatives in the behavior of real materials. J. Appl. Mech. **51**, 294–298 (1984)
3. Diethelm, K.: The Analysis of Fractional Differential Equations. Springer, New York (2004)
4. Hernandez, E., O'Regan, D., Balachandran, K.: On recent developments in the theory of abstract differential equations with fractional derivatives. Nonlinear Anal. Theor. Meth. Appl. **73**(10), 3462–3471 (2010)
5. Hilfer, R.: Applications of Fractional Calculus in Physics. World Scientific, River Edge, NJ (2000)
6. Li, F., N'Guérékata, G.M.: An existence result for neutral delay integrodifferential equation with fractional order and nonlocal conditions. Abstract Appl. Anal. **2011**, 20, Article ID 952782 (2011)

7. Lizama, C., N'Guérékata, G.M.: Mild solution for abstract fractional differential equations. Appl. Anal. DOI: 10.1080/00036811.2012.698003
8. Lv, L.L., Wang, J.R., Wei, W.: Existence and uniqueness results for fractional differential equations with boundary value conditions. Opuscula Nathematica. **31**(4), 1838–1843 (2011)
9. Mophou, G., N'Guérékata, G.M.: Existence of mild solutions of some semilinear neutral fractional functional evolution equations with infinite delay. Appl. Math. Comput. **216**(1), 61–69 (2010)
10. Mophou, G., N'Guérékata, G.M.: A note on semilinear fractional differential equations of neutral type with infinite delay. Adv. Differ. Equat. **2010**, 8, Article ID 674630 (2010)
11. Mophou, G., N'Guérékata, G.M.: Optimal control of a fractional diffusion equation with state constraints. Comput. Math. Appl. **62**, 1413–1426 (2011)
12. Mophou, G., N'Guérékata, G.M.: On a class of fractional differential equations in a Sobolev space. Appl. Anal. **91**(1), 15–34 (2012)
13. N'Guérékata, G. M.: A Cauchy problem for some fractional abstract differential equations with nonlocal conditions. Nonlinear Anal. Theor. Meth. Appl. **70**, 1873–1876 (2009)
14. Podlubny, I.: Fractional Differential Equations. Academic Press (1999)

Chapter 3
A Note on Nonautonomous Systems of Second-Order Differential Equations

Toka Diagana

Abstract The main objective of this paper is twofold. We first revisit the concept of doubly weighted pseudo-almost automorphy and discuss some additional properties of these functions. Next, we make extensive use of dichotomy tools to study and obtain the existence of doubly weighted pseudo-almost automorphic solutions to some nonautonomous second-order systems of differential equations. As an illustration, we will consider a nonresonance case for some scalar second-order systems of second-order differential equations.

Keywords Doubly weighted pseudo-almost automorphic • Nonresonance

Mathematics Subject Classification (2000): primary 42A75; secondary 34C27

3.1 Introduction

In Diagana [4] the concept of weighted pseudo-almost periodicity was introduced for the first time. Later on, Blot et al. [1] extended the notion of weighted pseudo-almost periodicity by introducing the notion of weighted pseudo-almost automorphy. More recently, in Diagana [5, 6], the notions of doubly weighted pseudo-almost periodicity and doubly weighed pseudo-almost automorphy were introduced, which generalize respectively the notions of weighted pseudo-almost periodicity and weighted pseudo-almost automorphy.

T. Diagana (✉)
Department of Mathematics, Howard University, 2441 6th Street N.W.,
Washington, DC 20059, USA
e-mail: tdiagana@howard.edu

B. Toni et al. (eds.), *Bridging Mathematics, Statistics, Engineering and Technology*,
Springer Proceedings in Mathematics & Statistics 24, DOI 10.1007/978-1-4614-4559-3_3,
© Springer Science+Business Media New York 2012

In this paper we first revisit the notion of doubly weighted pseudo-almost automorphy and discuss some additional properties of these classes of functions. Our next task consists of studying the existence of doubly weighted pseudo-almost automorphic solutions to

$$u''(t) + B(t)u'(t) + A(t)u(t) = f(t,u), \quad t \in \mathbb{R}, \tag{3.1}$$

where $A(t), B(t) : \mathbb{R}^n \mapsto \mathbb{R}^n$ are $n \times n$-square matrices with real coefficients and the function $f : \mathbb{R} \times \mathbb{R}^n \mapsto \mathbb{R}^n$ is jointly continuous and satisfies some additional conditions.

To illustrate the above-mentioned abstract case, we study a nonresonance case for the scalar second-order differential equation

$$u''(t) + b(t)u'(t) + a(t)u(t) = f(t,u), \quad t \in \mathbb{R}, \tag{3.2}$$

where the function $f : \mathbb{R} \times \mathbb{R} \mapsto \mathbb{R}$ is jointly continuous and satisfies some additional conditions, and the functions $a, b : \mathbb{R} \mapsto \mathbb{R}$ are almost automorphic and satisfy some additional conditions (see assumptions (H.5) and (H.6)).

To study Eq. (3.1), our strategy consists of rewriting it as a first-order system of differential equation in $\mathbb{R}^n \times \mathbb{R}^n$ involving a $2n \times 2n$ square matrix $\mathcal{A}(t)$. Indeed, if u is twice differentiable, if we denote the identity and the zero matrices of $\mathbb{R}^n$ respectively by $I_{\mathbb{R}^n}$ and $0_{\mathbb{R}^n}$, and setting

$$w := \begin{pmatrix} u \\ \\ u' \end{pmatrix} \in \mathbb{R}^n \times \mathbb{R}^n,$$

then Eq. (3.1) can be rewritten as follows:

$$w'(t) = \mathcal{A}(t)w(t) + F(t,w), \quad t \in \mathbb{R}, \tag{3.3}$$

where $\mathcal{A}(t)$ is the $2n \times 2n$ square matrix given by

$$\mathcal{A}(t) = \begin{pmatrix} 0_{\mathbb{R}^n} & I_{\mathbb{R}^n} \\ \\ -A(t) & -B(t) \end{pmatrix} \tag{3.4}$$

and the function F appearing in Eq. (3.3) is defined by

$$F(t,w) := \begin{pmatrix} 0 \\ \\ f(t,u) \end{pmatrix}.$$

3.1.1 Preliminaries

In this paper the real n-dimensional space $\mathbb{R}^n$ will be equipped with its natural Euclidean norm $|\cdot|$ defined for each $x = (x_1, x_2, \dots, x_n) \in \mathbb{R}^n$ by

$$|x| = \sqrt{x_1^2 + x_2^2 + \cdots + x_n^2}.$$

Let $BC(\mathbb{R}, \mathbb{R}^n)$ denote the collection of all-bounded continuous functions $f : \mathbb{R} \mapsto \mathbb{R}^n$. Clearly, $BC(\mathbb{R}, \mathbb{R}^n)$ equipped with the sup-norm defined by

$$\|f\|_\infty = \sup_{t \in \mathbb{R}} |f(t)|$$

is a Banach space.

The notation $\mathbb{U}$ will stand for the collection of (weight) functions $\rho : \mathbb{R} \mapsto (0, \infty)$, which are locally integrable upon $\mathbb{R}$ such that $\rho > 0$ almost everywhere. Now, if $\mu \in \mathbb{U}$ and if $r > 0$, we set $Q_r := [-r, r]$ and let

$$\mu(Q_r) := \int_{Q_r} \mu(t) \mathrm{d}t.$$

In this setting we are interested in weights $\mu \in \mathbb{U}$ for which $\lim_{r \to \infty} \mu(Q_r) = \infty$. Using the previous tools, we then define the set of weights $\mathbb{U}_\infty$ by

$$\mathbb{U}_\infty := \left\{ \mu \in \mathbb{U} : \lim_{r \to \infty} \mu(Q_r) = \infty \right\}.$$

We also need the following set of weights:

$$\mathbb{U}_b := \left\{ \mu \in \mathbb{U}_\infty : 0 < m_0 = \inf_{x \in \mathbb{R}} \mu(x) \le \sup_{x \in \mathbb{R}} \mu(x) = \mu_1 < \infty \right\}.$$

Definition 3.1.1. A function $f \in C(\mathbb{R}, \mathbb{R}^n)$ is said to be almost automorphic if for every sequence of real numbers $(s_n')_{n \in \mathbb{N}}$, there exists a subsequence $(s_n)_{n \in \mathbb{N}}$ such that

$$g(t) := \lim_{n \to \infty} f(t + s_n)$$

is well defined for each $t \in \mathbb{R}$, and

$$\lim_{n \to \infty} g(t - s_n) = f(t)$$

for each $t \in \mathbb{R}$.

Denote by $AA(\mathbb{R}^n)$ the collection of all almost automorphic functions $f : \mathbb{R} \mapsto \mathbb{R}^n$, which turns out to be a Banach space when equipped with the sup-norm $\|\cdot\|_\infty$.

Definition 3.1.2. A jointly continuous function $F : \mathbb{R} \times \mathbb{R}^m \mapsto \mathbb{R}^n$ is said to be almost automorphic in $t \in \mathbb{R}$ if $t \mapsto F(t,x)$ is almost automorphic for all $x \in K$ ($K \subset \mathbb{R}^m$ being any bounded subset). Equivalently, for every sequence of real numbers $(s'_n)_{n \in \mathbb{N}}$, there exists a subsequence $(s_n)_{n \in \mathbb{N}}$ such that

$$G(t,x) := \lim_{n \to \infty} F(t + s_n, x)$$

is well defined in $t \in \mathbb{R}$ and for each $x \in K$, and

$$\lim_{n \to \infty} G(t - s_n, x) = F(t,x)$$

for all $t \in \mathbb{R}$ and $x \in K$.

The collection of such functions will be denoted by $\mathrm{AA}(\mathbb{R}^m, \mathbb{R}^n)$.

For more on almost automorphic functions we refer the reader to the book of N'Guérékata [8].

If $\mu, \nu \in \mathbb{U}_\infty$, we define

$$\mathrm{PAP}_0(\mathbb{R}^n, \mu, \nu) := \left\{ f \in BC(\mathbb{R}, \mathbb{R}^n) : \ \lim_{T \to \infty} \frac{1}{\mu(Q_T)} \int_{Q_T} \|f(\sigma)\| \, \nu(\sigma) \, \mathrm{d}\sigma = 0 \right\}.$$

Similarly, we define $\mathrm{PAP}_0(\mathbb{R}^m, \mathbb{R}^n, \mu, \nu)$ as the collection of jointly continuous functions $F : \mathbb{R} \times \mathbb{R}^m \mapsto \mathbb{R}^n$ such that $F(\cdot, y)$ is bounded for each $y \in \mathbb{R}^m$ and

$$\lim_{T \to \infty} \frac{1}{\mu(Q_T)} \left\{ \int_{Q_T} |F(s,y)| \, \nu(s) \, \mathrm{d}s \right\} = 0$$

uniformly in $y \in \mathbb{R}^m$.

The space $\mathrm{PAP}_0(\mathbb{R}^n, \mu, \mu)$ is denoted by $\mathrm{PAP}_0(\mathbb{R}^n, \mu)$. Similarly, $\mathrm{PAP}_0(\mathbb{R}^m, \mathbb{R}^n, \mu, \mu)$ is denoted by $\mathrm{PAP}_0(\mathbb{R}^m, \mathbb{R}^n, \mu)$.

Definition 3.1.3 ([9]). A continuous function $F(t,s) : \mathbb{R} \times \mathbb{R} \mapsto \mathbb{R}^n$ is called bi-almost automorphic if for every sequence of real numbers (s_m) we can extract a subsequence (s_n) such that

$$G(t,s) := \lim_{n \to \infty} F(t + s_n, s + s_n)$$

is well defined for all $t, s \in \mathbb{R}$, and

$$\lim_{n \to \infty} G(t - s_n, s - s_n) = F(t,s)$$

for all $t, s \in \mathbb{R}$. The collection of such functions is denoted $\mathrm{bAA}(\mathbb{R} \times \mathbb{R}, \mathbb{R}^n)$.

The following definitions of doubly weighted pseudo-almost automorphy are due to Diagana [5, 6].

Definition 3.1.4. Let $\mu \in \mathbb{U}_\infty$ and $v \in \mathbb{U}_\infty$. A function $f \in C(\mathbb{R}, \mathbb{R}^n)$ is called doubly weighted pseudo-almost automorphic if it can be expressed as $f = g + \phi$, where $g \in AA(\mathbb{R}^n)$ and $\phi \in PAP_0(\mathbb{R}^n, \mu, v)$. The collection of such functions will be denoted by $PAP(\mathbb{R}^n, \mu, v)$.

Definition 3.1.5. Let $\mu, v \in \mathbb{U}_\infty$. A function $f \in C(\mathbb{R} \times \mathbb{R}^m, \mathbb{R}^n)$ is called doubly weighted pseudo-almost automorphic if it can be expressed as $F = G + \Phi$, where $G \in AA(\mathbb{R}^m, \mathbb{R}^n)$ and $\Phi \in PAP_0(\mathbb{R}^m, \mathbb{R}^n, \mu, v)$. The collection of such functions will be denoted by $PAA(\mathbb{R}^m, \mathbb{R}^n, \mu, v)$.

Let $\mu, v \in \mathbb{U}_\infty$. According to Diagana [6, Theorem 2.16], if $PAP_0(\mathbb{R}^n, \mu, v)$ is translation invariant and if

$$\inf_{r>0} \frac{v(Q_r)}{\mu(Q_r)} > 0, \tag{3.5}$$

then the decomposition of doubly weighted pseudo-almost automorphic functions is unique.

Let $\mathbb{W}_\infty$ be the collection of all weights $\rho \in \mathbb{U}_\infty$ such that

$$\overline{\lim_{|t| \to \infty}} \frac{\rho(t + \tau)}{\rho(t)} < \infty, \ \forall \tau \in \mathbb{R}.$$

According to a recent paper by Ji and Zhang [7], if $\rho \in \mathbb{W}_\infty$, then $PAP_0(\mathbb{R}^n, \rho)$ is translation invariant. Similarly, it can be easily seen that if $\mu, v \in \mathbb{W}_\infty$, then $PAP_0(\mathbb{R}^n, \mu, v)$ is translation invariant, too. The previous discussion on the uniqueness of the decomposition of doubly weighted pseudo-almost automorphic functions can be formulated as follows:

Proposition 3.1.6. *If $\mu, v \in \mathbb{W}_\infty$ and if (3.5) holds, then the decomposition of doubly weighted pseudo-almost automorphic functions is unique, that is,*

$$PAA(\mathbb{R}^n, \mu, v) = AA(\mathbb{R}^n) \oplus PAP_0(\mathbb{R}^n, \mu, v).$$

We need the following composition result for doubly weighted pseudo-almost automorphic functions, which was obtained by Diagana [5].

Theorem 3.1.7 ([5]). *Let $\mu, v \in \mathbb{U}_\infty$ and let $f \in PAA(\mathbb{R}^m, \mathbb{R}^n, \mu, v)$ satisfying the Lipschitz condition; there exists $L \geq 0$ such that*

$$|f(t, u) - f(t, v)| \leq L|u - v| \quad \text{for all } u, v \in \mathbb{R}^m, \ t \in \mathbb{R}.$$

If $h \in PAA(\mathbb{R}^m, \mu, v)$, then $f(\cdot, h(\cdot)) \in PAA(\mathbb{R}^n, \mu, v)$.

3.2 Existence of Doubly Weighted Pseudo-Almost Automorphic Solutions

In the rest of the paper we suppose that $\mu, \nu \in \mathbb{W}_\infty$ and that Eq. (3.5) holds. Let $\{A(t)\}_{t \in \mathbb{R}}$ be an $n \times n$ square matrix and consider the first-order system of differential equations given by

$$z'(t) = A(t)z(t) + g(t), \quad t \in \mathbb{R} \tag{3.6}$$

and its corresponding homogeneous equation

$$z'(t) = A(t)z(t), \quad t \in \mathbb{R} \tag{3.7}$$

where $g : \mathbb{R} \mapsto \mathbb{R}^n$ is continuous.

Definition 3.2.1 ([2]). The homogeneous equation (3.7) is said to be to have an exponential dichotomy if there exist a projection P and the constants $K, \delta > 0$ such that

(i) $\|X(t)PX^{-1}(s)\| \le Ke^{-\delta(t-s)}$ for all $t, s \in \mathbb{R}$ and $t \ge s$; and
(ii) $\|X(t)QX^{-1}(s)\| \le Ke^{-\delta(s-t)}$ for all $t, s \in \mathbb{R}$ and $t \le s$.

where $Q = I - P$ and $X(t)$ is a fundamental solution to Eq. (3.10) satisfying $X(0) = I$.

If Eq. (3.10) has an exponential dichotomy, we then define

$$\Gamma(t,s) = \begin{cases} X(t)PX^{-1}(s) & \text{if } t \ge s, \\ X(t)QX^{-1}(s) & \text{if } s \ge t. \end{cases}$$

It can be easily seen that

$$\|\Gamma(t,s)\| \le \begin{cases} Ke^{-\delta(t-s)} & \text{if } t \ge s, \\ Ke^{-\delta(s-t)} & \text{if } s \ge t. \end{cases}$$

Our setting requires the following additional assumptions:

(H.1) $g \in PAA(\mathbb{R}^n, \mu, \nu)$
(H.2) $\Gamma(t,s)u \in bAA(\mathbb{R} \times \mathbb{R}, \mathbb{R}^n)$ uniformly for all u in any bounded subset of $\mathbb{R}^n$

We have

Theorem 3.2.2. *If Eq. (3.7) has exponential dichotomy and if assumptions (H.1) and (H.2) hold, then Eq. (3.6) has a unique doubly weighted pseudo-almost automorphic solution.*

Proof. The proof is slightly similar to the one given in a recent paper by Diagana and Nelson [3]. However, for the sake of clarity, we reproduce it here. Indeed, let $X(t)$ be a fundamental solution to Eq. (3.7) satisfying $X(0) = I$ and suppose there exist a projection P and the constants $K, \delta > 0$ such that

$$\|X(t)PX^{-1}(s)\| \le Ke^{-\delta(t-s)} \tag{3.8}$$

for all $t, s \in \mathbb{R}$ and $t \geq s$; and

$$\|X(t)QX^{-1}(s)\| \leq Ke^{-\delta(s-t)} \tag{3.9}$$

for all $t, s \in \mathbb{R}$ and $t \leq s$, where $Q = I - P$.

Set $U(t,s) = X(t)PX^{-1}(s)$ for $t \geq s$ and $\widetilde{U}(t,s) = X(t)QX^{-1}(s)$ for $t \leq s$. According to Coppel [2], the only bounded solution to Eq. (3.6) is given by

$$z(t) = \int_{-\infty}^{t} U(t,s)g(s)ds - \int_{t}^{\infty} \widetilde{U}(t,s)g(s)ds.$$

Let $g = g_1 + g_2 \in \mathrm{PAA}(\mathbb{R}^n)$ where $g_1 \in \mathrm{AA}(\mathbb{R}^n)$ and $g_2 \in \mathrm{PAP}_0(\mathbb{R}^n)$. We also set

$$Sg_j(t) := \int_{-\infty}^{t} U(t,s)g_j(s)ds \text{ and } Rg_j(t) := \int_{t}^{\infty} \widetilde{U}(t,s)g_j(s)ds \quad \text{for } j = 1, 2.$$

Let us show that $Sg_1 \in \mathrm{AA}(\mathbb{R}^n)$. Indeed, since $g_1 \in \mathrm{AA}(\mathbb{R}^n)$, for every sequence of real numbers $(\tau'_n)_{n \in \mathbb{N}}$ there exists a subsequence $(\tau_n)_{n \in \mathbb{N}}$ such that

$$h_1(t) := \lim_{n \to \infty} g_1(t + \tau_n)$$

is well defined for each $t \in \mathbb{R}$, and

$$\lim_{n \to \infty} h_1(t - \tau_n) = g_1(t)$$

for each $t \in \mathbb{R}$.

We have

$$Sg_1(t + \tau_n) - Sh_1(t) = \int_{-\infty}^{t+\tau_n} U(t + \tau_n, s)g_1(s)ds - \int_{-\infty}^{t} U(t,s)h_1(s)ds$$

$$= \int_{-\infty}^{t} U(t + \tau, s + \tau_n)g_1(s + \tau_n)ds - \int_{-\infty}^{t} U(t,s)h_1(s)ds$$

$$= \int_{-\infty}^{t} U(t + \tau_n, s + \tau_n)(g_1(s + \tau_n) - h_1(s))ds$$

$$+ \int_{-\infty}^{t} (U(t + \tau_n, s + \tau_n) - U(t,s))h_1(s)ds.$$

Using Eq. (3.8) and the Lebesgue-Dominated Convergence theorem, one can easily see that

$$\left| \int_{-\infty}^{t} U(t + \tau_n, s + \tau_n)(g_1(s + \tau_n) - h_1(s))ds \right| \to 0 \text{ as } n \to \infty, t \in \mathbb{R}.$$

Similarly, using (H.2) it follows that

$$\left| \int_{-\infty}^{t} (U(t + \tau_n, s + \tau_n) - U(t,s)) h_1(s) ds \right| \to 0 \quad \text{as } n \to \infty, \, t \in \mathbb{R}.$$

Therefore,

$$Sh_1(t) = \lim_{n \to \infty} Sg_1(t + \tau_n), \quad t \in \mathbb{R}.$$

Using ideas similar to the previous ones, one can easily see that

$$Sg_1(t) = \lim_{n \to \infty} Sh_1(t - \tau_n), \quad t \in \mathbb{R}.$$

Similarly, using again Eq. (3.8) it follows that

$$\lim_{r \to \infty} \frac{1}{v(Q_r)} \int_{Q_r} |(Sg_2)(t)| \mu(t) dt \leq \lim_{r \to \infty} \frac{K}{v(Q_r)} \int_{Q_r} \int_0^{+\infty} e^{-\delta s} |g_2(t - s)| \mu(t) ds dt$$

$$\leq \lim_{r \to \infty} K \int_0^{+\infty} e^{-\delta s} \frac{1}{v(Q_r)} \int_{Q_r} |g_2(t - s)| \mu(t) dt ds.$$

Set

$$\Gamma_s(r) = \frac{1}{v(Q_r)} \int_{Q_r} |g_2(t - s)| \mu(t) dt.$$

Since $PAP_0(\mathbb{R}^n, \mu, v)$ is translation invariant it follows that $t \mapsto g_2(t - s)$ belongs to $PAP_0(\mathbb{R}^n, \mu, v)$ for each $s \in \mathbb{R}$, and hence

$$\lim_{r \to \infty} \frac{1}{v(Q_r)} \int_{Q_r} |g_2(t - s)| \mu(t) dt = 0$$

for each $s \in \mathbb{R}$. One completes the proof by using the well-known Lebesgue-Dominated Convergence theorem and the fact $\Gamma_s(r) \mapsto 0$ as $r \to \infty$ for each $s \in \mathbb{R}$.

The proof for R is similar to that of S and hence omitted. For R, one makes use of Eq. (3.9) rather than Eq. (3.8).   □

In order to apply the previous result to Eq. (3.1) and then to Eq. (3.3), we need to make the following additional assumptions:

(H.3) There exists $L > 0$ such that

$$|f(t, u) - f(t, v)| \leq L|u - v| \quad \text{for all } u, v \in \mathbb{R}^n, \, t \in \mathbb{R}.$$

(H.4) $f \in PAA(\mathbb{R}^n, \mathbb{R}^n, \mu, v)$.

Theorem 3.2.3. *If Eq. (3.10) has exponential dichotomy and if assumptions* (H.2), (H.3), (H.4) *hold, then Eq. (3.1) has a unique doubly weighted pseudo-almost automorphic solution whenever L is small enough.*

Proof. Define the nonlinear integral operator Γ defined by

$$(\Lambda u)(t) = \int_{-\infty}^{t} U(t,s) f(s, u(s)) \, ds - \int_{t}^{\infty} \tilde{U}(t,s) f(s, u(s)) \, ds.$$

Let $u \in \text{PAA}(\mathbb{R}^n, \mathbb{R}^n, \mu, \nu)$. Using assumptions (H.3) and (H.4) and Theorem 3.1.7 it follows that $g(s) := f(s, u(s))$ belongs to $\text{PAA}(\mathbb{R}^n, \mu, \nu)$. Next, one can easily show that

$$\Lambda(\text{PAA}(\mathbb{R}^n, \mu, \nu)) \subset \text{PAA}(\mathbb{R}^n, \mu, \nu).$$

Now if $u, v \in \text{PAA}(\mathbb{R}^n, \mu, \nu)$ are arbitrarily chosen elements, then

$$\|\Lambda(u) - \Lambda(v)\|_{\infty} \leq KL\delta^{-1} \|u - v\|_{\infty}.$$

Therefore, if L is small enough, that is, $L < K^{-1}\delta$, then Λ has a unique fixed point which obviously is the unique solution to

$$u'(t) = A(t)u(t) + f(t, u), \quad t \in \mathbb{R}.$$

Clearly, under the same assumptions it follows that Eq. (3.3) has a unique doubly weighted pseudo-almost automorphic solution given by

$$t \to z(t) := \begin{pmatrix} u(t) \\ \\ u'(t) \end{pmatrix}.$$

Therefore, Eq. (3.1) has a unique doubly weighted pseudo-almost automorphic solution u whenever L is small enough. In fact, the solution u belongs to a smaller space than $\text{PAA}(\mathbb{R}^n, \mu, \nu)$, that is, $u \in \text{PAA}^{(1)}(\mathbb{R}^n, \mu, \nu) \subset \text{PAA}(\mathbb{R}^n, \mu, \nu)$, where $\text{PAA}^{(1)}(\mathbb{R}^n, \mu, \nu)$ stands for the space of all $C^{(1)}$-doubly-weight pseudo-almost automorphic functions consisting of all functions $\varphi \in \text{PAA}(\mathbb{R}^n, \mu, \nu)$ such that $\varphi' \in \text{PAA}(\mathbb{R}^n, \mu, \nu)$. $\qquad \square$

3.3 A Nonresonance Case

Let $n = 1$. In order to illustrate Theorem 3.2.3, we will study a nonresonance case (see (H.6)) for the scalar second-order systems differential equations given by Eq. (3.2). For that, we let $A(t) = a(t)$ and $B(t) = b(t)$, and suppose that both $a, b : \mathbb{R} \mapsto \mathbb{R}$ are almost automorphic and satisfy the following additional assumptions:

(H.5) There exist $a_0, b_0 > 0$ such that

$$\inf_{t \in \mathbb{R}} a(t) = a_0 \quad \text{and} \quad \inf_{t \in \mathbb{R}} b(t) = b_0.$$

(H.6) $b(t) \neq 2\sqrt{a(t)}$ for all $t \in \mathbb{R}$.

Now

$$\mathcal{A}(t) = \begin{pmatrix} 0 & 1 \\ -a(t) & -b(t) \end{pmatrix}$$

which yields $P_t(\lambda) = \det(\mathcal{A}(t) - \lambda I_{\mathbb{R}^2}) = \lambda^2 + b(t)\lambda + a(t)$ for all $t \in \mathbb{R}$.

Let $D(t) = b^2(t) - 4a(t)$ for all $t \in \mathbb{R}$. Clearly, (H.6) yields either $D(t) > 0$ or $D(t) < 0$ for all $t \in \mathbb{R}$.

- If $D(t) > 0$ for all $t \in \mathbb{R}$ and if assumptions (H.5) and (H.6) hold, then eigenvalues of $\mathcal{A}(t)$ are given by

$$\lambda_1(t) = \frac{-b(t) + \sqrt{b^2(t) - 4a(t)}}{2} \quad \text{and} \quad \lambda_2(t) = \frac{-b(t) - \sqrt{b^2(t) - 4a(t)}}{2}.$$

It is then easy to see that $\lambda_1(t)$, $\lambda_2(t) < 0$ for all $t \in \mathbb{R}$.

- If $D(t) < 0$ for all $t \in \mathbb{R}$ and if (H.5) and (H.6) hold, then eigenvalues of $\mathcal{A}(t)$ are given by

$$\lambda_1(t) = \frac{-b(t) + i\sqrt{4a(t) - b^2(t)}}{2} \quad \text{and} \quad \lambda_2(t) = \frac{-b(t) - i\sqrt{4a(t) - b^2(t)}}{2}.$$

It is then easy to see that $\mathfrak{Re}\lambda_1(t)$, $\mathfrak{Re}\lambda_2(t) < 0$ for all $t \in \mathbb{R}$.

Thus under assumptions (H.5) and (H.6), one has $\mathfrak{Re}(\lambda_k(t)) < 0$ for all $t \in \mathbb{R}$ for $k = 1, 2$ (a nonresonance case). It follows that there exists $\omega > 0$ and $M > 0$ such that

$$\|e^{s\mathcal{A}(t)}\| \leq Me^{-\omega s}, \quad s \geq 0$$

which yields

$$w'(t) = \mathcal{A}(t)w(t), \quad t \in \mathbb{R} \tag{3.10}$$

has exponential dichotomy with projection $P = I_{\mathbb{R}^2}$.

Using similar techniques as in the proof of Theorem 3.2.3 we obtain the following:

Theorem 3.3.1. *Under assumptions (H.3)–(H.6), then Eq. (3.2) has a unique doubly weighted pseudo-almost automorphic solution whenever L is small enough.*

References

1. Blot, J., Mophou, G., N'Guérékata, G., Pennequin, D.: Weighted pseudo almost automorphic functions and applications to abstract differential equations. Nonlinear Anal. **71**(3-4), 903–909 (2009)

2. Coppel, W.A.: Dichotomies in stability theory. Lecture Notes in Mathematics, vol. 629. Springer-Verlag, Berlin-New York (1978)
3. Diagana, T., Nelson, V.: $C^{(n)}$-Pseudo almost automorphy and its applications to some higher-order differential equations. Nonlinear Stud. **19**(3) (2012)
4. Diagana, T. Weighted pseudo-almost periodic functions and applications. C. R. Acad. Sci. Paris, Ser I **343**(10), 643–646 (2006)
5. Diagana, T.: Doubly-weighted pseudo almost periodic functions. African Diaspora J. Math. **12**(1), 121–136 (2011)
6. Diagana, T.: Existence of doubly-weighted pseudo almost periodic solutions to non-autonomous differential equations. Electron. J. Differ. Equat. **28**, 15 (2011)
7. Ji, D., Zhang, C.: Translation invariance of weighted pseudo almost periodic functions and related issues. J. Math. Anal. Appl. **391**, 350–362 (2012)
8. N'Guérékata, G.M.: Almost Automorphic Functions and Almost Periodic Functions in Abstract Spaces. Kluwer Academic / Plenum Publishers, New York (2001)
9. Xiao, T. J., Zhu, X-X., Liang, J.: Pseudo-almost automorphic mild solutions to nonautonomous differential equations and applications. Nonlinear Anal. **70**(11), 4079–4085 (2009)

Chapter 4
How the Talmud Divides an Estate Among Creditors

Stephen Schecter

Abstract The Talmud gives examples of how to divide an estate among creditors when the debts total more than the estate, but it is not clear what the algorithm is. We describe the solution of this problem by Aumann, Maschler, and Kaminski, and its relation to game theory.

Keywords Bankruptcy • Hydraulic rationing • Nucleolus

4.1 Introduction

A man dies leaving an estate that is too small to pay his debts. How much should each creditor get?

The Babylonian Talmud, a compendium of Jewish law that dates back 1,800 years, gives the following example. Creditor 1 is owed 100, Creditor 2 is owed 200, and Creditor 3 is owed 300.

1. If the estate is 100, each creditor gets 33 1/3.
2. If the estate is 200, Creditor 1 gets 50, Creditors 2 and 3 get 75 each.
3. If the estate is 300, Creditor 1 gets 50, Creditor 2 gets 100, Creditor 3 gets 150.

A literature stretching across 1,500 years deals with the question: what algorithm is the Talmud using? Of course, as in any legal system, the answer must be based on the system's principles and precedents.

The problem was convincingly solved by two mathematicians at Hebrew University of Jerusalem, Robert Aumann and Michael Maschler, in the 1980s [1]. Later Marek Kaminski, a political scientist now at the University of California,

S. Schecter (✉)
Department of Mathematics, North Carolina State University, Raleigh, NC 27695, USA
e-mail: schecter@math.ncsu.edu

B. Toni et al. (eds.), *Bridging Mathematics, Statistics, Engineering and Technology*,
Springer Proceedings in Mathematics & Statistics 24, DOI 10.1007/978-1-4614-4559-3_4,
© Springer Science+Business Media New York 2012

Irvine, showed how, given the sizes of the debts, one can construct special-purpose glassware so that when an amount of liquid equal to the size of the estate is poured in, it will divide itself in the correct way [3].

The goal of this paper is to explain the Aumann–Maschler–Kaminski solution and its relation to game theory.

The estate-division problem is related to bankruptcy, since the same issue arises in dividing the assets of a bankrupt person or corporation among creditors. As we shall see, the Aumann–Maschler–Kaminski solution is related to the Talmud's view of this and other situations in which money is owed.

My interest in the estate-division problem grew out of a game theory course for undergraduates that I teach. When Aumann was awarded the Nobel Prize in Economics in 2005 for work in game theory, I read some of what was written about him. Most of his work is rather technical. Perhaps for that reason, journalists writing about Aumann tended to move quickly to the fact that he had solved an old problem from the Talmud, apparently because this was thought to be of general interest. I looked into what the problem was and found a story that opens in many directions.

Aumann learned of the Talmud's estate-division problem in 1980 or 1981 from his son Shlomo, who was studying at a Talmudic academy in Jerusalem and pointed his father to the relevant passage. Shlomo Aumann was killed in 1982 while serving in the Israeli army.

4.2   What is the Talmud?

The Talmud (more precisely, the Babylonian Talmud) consists of:

- The Mishna (c. 200 CE), a written compendium of Judaism's Oral Law
- The Gemara (c. 500 CE), a record of discussions by rabbis about the Mishna

It is divided into 60 tractates, or books. The first printed version appeared in Italy around 1,520, some 85 years after Gutenberg invented the printing press. A modern edition with English translation occupies 73 volumes.

Figure 4.1 shows a page of Talmud from a modern edition. The unusual form of the page dates back to the earliest printings. The header at the top of the page gives the tractate, in this case Megillah, or scroll, meaning the scroll of Esther; this tractate deals with laws concerning the reading of the Book of Esther at the holiday of Purim. The header also gives the chapter number and name and the page number. Modern references to the Talmud give just the tractate and page number, followed by the letter a or b, meaning front or back of the page. The central block of text is portions of Mishna and the related Gemara, separated by colons. These texts are often fairly obscure; the Gemara is often in the form of notes on a discussion. The Mishna is in Hebrew, the Gemara in Aramaic. Commentary by Rashi (a French rabbi, 1040–1105) wraps around the central block at the upper right. The Talmud is considered to be largely incomprehensible without Rashi's commentary. Wrapping around the central block at the left is commentary by Rashi's successors in the

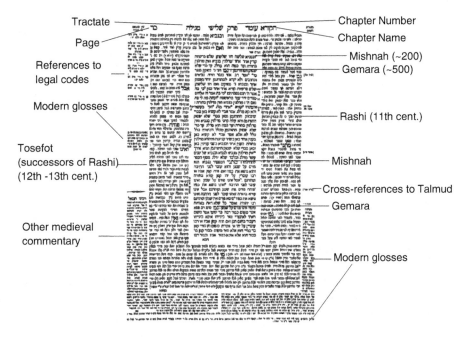

Fig. 4.1 A page of Talmud

twelfth and thirteenth centuries. Outside these texts are more recent commentary, cross-references to related pages of Talmud, and references to codifications of Talmudic law.

The word "Mishna" is also used to refer to single portion of Mishna on a page of Talmud.

4.3 A Problem from the Talmud

A man dies leaving

- An estate of size e
- Debts to creditors $1, \ldots, n$ of $d_1, \ldots, d_n$
- $e < d_1 + \cdots + d_n$

How much should each creditor get?

A Mishna (Tractate Ketubot 93a) gives the answer described in the Introduction. ("Ketubot" is the plural of ketubah, which means marriage contract. This tractate includes laws about marriage and legal and financial aspects of family life.) Alfasi (a Moroccan rabbi, 1013–1103) wrote, "My predecessors discussed this Mishna and its Gemara at length, and were unable to make sense of it." Aumann and Maschler write in [1]: "Over two millennia, this Mishna has spawned a large literature. Many authorities disagree with it outright. Others attribute the figures to special circumstances, not made explicit in the Mishna. A few have attempted direct rationalizations of the figures as such, mostly with little success. One modern scholar, exasperated by his inability to make sense of the text, suggested errors in transcription. In brief, the passage is notoriously difficult."

An *estate-division problem* is a pair $(e, (d_1, \ldots, d_n))$ with the following properties:

1. $0 < d_1 \le d_2 \le \cdots \le d_n$.
2. Let $d = d_1 + \cdots + d_n$. Then $0 < e < d$.

A *division* of the estate is an n-tuple $(x_1, \ldots, x_n)$ with $0 \le x_i$ for all i and $x_1 + \cdots + x_n = e$.

Here are some ideas about how an estate should be divided.

Proportional Division. Compute the fraction of the total debt that is owed each creditor, and assign her that fraction of the estate. (Following Aumann and Maschler, we use the pronoun "her" because in the Talmudic example, the creditors are women.) In other words, assign to creditor i the amount $\frac{d_i}{d}e$. Secular legal systems typically follow this idea, which treats each dollar of debt as equally worthy of payment. To most of us this approach seems obviously correct. Our Mishna appears to use this idea when $e = 300$.

Equal Division of Gains. Assign to each creditor the amount $\frac{e}{n}$. This method treats each *creditor* as equally worthy of payment. Our Mishna appears to use this idea when $e = 100$. Equal Division of Gains is not sensible if $d_1 < \frac{e}{n}$, since the first creditor (at least) will be paid more than she is owed. In other words, Equal Division of Gains is not sensible for large estates.

Constrained Equal Division of Gains. Give each creditor the same amount, but don't give any creditor more than her claim. In other words, choose a number a such that
$$\min(d_1, a) + \min(d_2, a) + \cdots + \min(d_n, a) = e.$$
Then assign to creditor i the amount $\min(d_i, a)$. The number a exists and is unique because for fixed $(d_1, \ldots, d_n)$, the left-hand side is a function of a that maps the interval $[0, e]$ onto itself and is strictly increasing on this interval. This rule was adopted by Maimonides (1135–1204, born in Spain, worked in Morocco and Egypt) in his codification of Talmudic law, the Mishneh Torah, which is still considered canonical. Maimonides' choice is inconsistent with our Mishna (it produces equal division in all our cases).

Equal Division of Losses. Make each creditor take the same loss. The total loss to the creditors is $d - e$, so assign to creditor i the amount $d_i - (d - e)/n$. This is not sensible if $d_1 < (d - e)/n$, since Creditor 1's portion of the estate would be negative. In other words, Equal Division of Losses not sensible for small estates.

Constrained Equal Division of Losses. Make each creditor take the same loss, but don't make any creditor lose more than her claim.

The principle that losses should be shared equally was used by Maimonides in a different context. Suppose that at an auction, n bidders bid amounts $b_1 < b_2 < \cdots < b_{n-1} < b_n$. The object is sold to the highest bidder for the price b_n. If for some reason the highest bidder reneges, the object is sold to the second-highest bidder for b_{n-1}. The highest bidder's reneging has cost the seller the difference between the two bids, $b_n - b_{n-1}$. Maimonides says that the highest bidder is obligated to pay the seller this amount. Now suppose that all n bidders renege. This costs the seller b_n, the amount he should have sold the object for. Maimonides says that each bidder must pay the amount $\frac{b_n}{n}$ to the seller. The bidders lose equal amounts to cover what the seller should have gained.

Actually, Maimonides just gives a numerical example. In his example, equal payments by each bidder would result from either the Equal Division of Losses principle or the Constrained Equal Division of Losses principle. We can guess, based on Maimonides' adoption of Constrained Equal Division of Gains for division of estates, that what he had in mind was Constrained Equal Division of Losses.

4.4 The Aumann–Maschler Solution

Aumann and Maschler's solution to the Talmud's estate-division problem was based on another Mishna and an issue dealt with in Gemara.

4.4.1 The Contested Garment Rule

The relevant Mishna is from Tractate Bava Metzia 2a: "Two hold a garment; one claims it all, the other claims half. Then the one is awarded three-fourths, the other one-fourth."

("Bava Metzia" means middle gate. It deals with civil law, including property law. The name refers to the gates of a city, where markets were located.)

Rashi explains the reasoning. The one who claims half concedes that half belongs to the other. Therefore only half is in dispute. It is split equally between the two claimants.

Alfasi, in his commentary mentioned earlier, says that Rabbi Hai Gaon suggested without giving details that the Mishna about estate division should be explained using Bava Mezia 2a. Hai Gaon (939–1083) worked in what is today the Iraqi city of Falujah.

The second relevant passage is from Gemara in Tractate Yevamot 38a. ("Yevamot" is the plural of yibum, levirate marriage, i.e., the requirement that a widow marry her deceased husband's brother. This requirement is found in Deuteronomy 25:5–6.) It deals with a rather complicated family tree, illustrated in Fig. 4.2.

Mr. B dies childless. His widow, as is required, marries his brother, C. C already has two sons, c_1 and c_2, by his first wife. Eight months later B's widow gives birth

Fig. 4.2 W_B is B's widow.
W_C is C's first wife

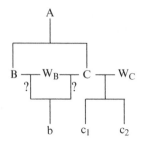

to a son, b, whose father is therefore doubtful. Next C dies. Finally, A, the father of B and C dies. How is A's estate to be divided among his grandchildren b, c_1, and c_2?

Young Mr. b says: Half of A's estate goes to A's son B and half to A's son C. I am B's only son, so I get his half. C's half should be divided between c_1 and c_2.

C's sons c_1 and c_2 say: B had no children, and C had three sons. Therefore the entire estate goes to C, and then is divided equally among the three grandchildren.

The Talmud's decision: c_1 and c_2 are treated as one claimant, b as another. The half of the estate that b concedes is not his goes to c_1 and c_2. The third of the estate that c_1 and c_2 concede is not theirs goes to b. The remainder of the estate, 1/6, is split equally: 1/12 to c_1 and c_2, 1/12 to b. Thus b gets 5/12 of the estate, and c_1 and c_2 get 7/12 to split.

Neither passage of Talmud treats a situation exactly analogous to an estate with creditors: there all claims are valid, whereas in these two passages, both claims cannot be valid. Nevertheless, applied to an estate with two creditors, we get:

Contested Garment Rule. Consider an estate-division problem with two creditors: $0 < d_1 \leq d_2$, $0 < e < d_1 + d_2$. Creditor 2 concedes $\max(e - d_2, 0)$ to Creditor 1. Creditor 1 concedes $\max(e - d_1, 0)$ to Creditor 2. The remainder of the estate, $e - \max(e - d_1, 0) - \max(e - d_2, 0)$, is divided equally. Thus Creditor 1 receives

$$\max(e - d_2, 0) + \frac{1}{2}(e - \max(e - d_1, 0) - \max(e - d_2, 0)).$$

Creditor 2 receives

$$\max(e - d_1, 0) + \frac{1}{2}(e - \max(e - d_1, 0) - \max(e - d_2, 0)).$$

4.4.2 Aumann and Maschler's Theorem

Here's our Mishna again. There are three debts, $d_1 = 100$, $d_2 = 200$, $d_3 = 300$.

1. If $e = 100$, each creditor gets 33 1/3.
2. If $e = 200$, Creditor 1 gets 50, Creditors 2 and 3 get 75 each.
3. If $e = 300$, Creditor 1 gets 50, Creditor 2 gets 100, Creditor 3 gets 150.

Fig. 4.3 Division of the
estate when $e \leq d_1$

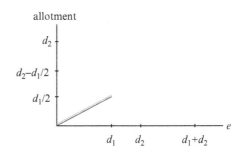

Aumann and Maschler observed that each of these divisions is consistent with
the Contested Garment Rule in the following sense. If any two creditors use the
Contested Garment Rule to split the amount they were jointly awarded, each will
get the amount she was actually awarded.

For example, look at the division with $e = 200$. Creditors 1 and 2 between them
are awarded 125. Consider an estate of size 125 with two claims on it, $d_1 = 100$
and $d_2 = 200$ (these were the original claims of Creditors 1 and 2). According to
the Contested Garment Rule, Creditor 1 concedes 25 to Creditor 2, and Creditor
2 concedes nothing to Creditor 1. The remaining 100 is split equally between the
two. Thus Creditor 1 gets 50 and Creditor 2 gets 75. These are the amounts that the
Mishna awarded them.

In an estate-division problem $(e, (d_1, \ldots, d_n))$, a division $(x_1, \ldots, x_n)$ of the estate
is *consistent with the Contested Garment Rule* if, for each pair (i, j), (x_i, x_j) is
exactly the division produced by the Contested Garment Rule applied to an estate
of size $x_i + x_j$ with debts d_i and d_j.

Aumann and Maschler proved:

Theorem 4.4.1 (Aumann–Maschler). *In any estate-division problem, there is
exactly one division of the estate that is consistent with the Contested Garment Rule.*

This is the division that the Talmud presumably has in mind. Aumann and
Maschler's proof was algebraic. We won't give it, since a nicer one appeared some
years later.

4.4.3   Another look at the Contested Garment Rule

Let's look more closely at the Contested Garment Rule for an estate with two
creditors.

1. If $e \leq d_1$, neither creditor concedes anything to the other, so the estate is split
   equally: each creditor gets $\frac{e}{2}$. Each additional dollar of estate value produces an
   equal gain for each creditor. See Fig. 4.3.

Fig. 4.4 Division of the
estate when $e \leq d_2$

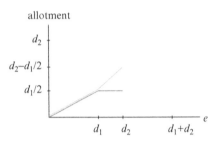

Fig. 4.5 Division of the
estate for all $e \leq d_1 + d_2$

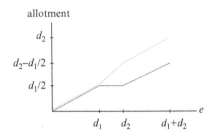

2. If $d_1 < e \leq d_2$, $e - d_1$ is conceded by Creditor 1 to Creditor 2, nothing is conceded
   by Creditor 2 to Creditor 1, and the remainder, d_1, is split equally.

$$\text{Creditor 1: } \frac{d_1}{2}. \qquad \text{Creditor 2: } (e - d_1) + \frac{d_1}{2}.$$

See Fig. 4.4. When $e = d_1$, the estate is split equally, so each Creditor has a gain
of $\frac{d_1}{2}$. Thereafter each additional dollar of estate value goes to Creditor 2. When
e reaches d_2, Creditor 1 gets $\frac{d_1}{2}$ and Creditor 2 gets $d_2 - \frac{d_1}{2}$, so each creditor has a
loss of $\frac{d_1}{2}$ relative to the debt she is owed. Previously Creditor 2's loss was larger.

3. If $d_2 < e < d_1 + d_2$, $e - d_1$ is conceded by Creditor 1 to Creditor 2, $e - d_2$ is
   conceded by Creditor 2 to Creditor 1, and the remainder, $e - (e - d_1) - (e - d_2) = d_1 + d_2 - e$, is split equally.

$$\text{Creditor 1: } e - d_2 + \frac{1}{2}(d_1 + d_2 - e) = \frac{d_1}{2} + \frac{1}{2}(e - d_2).$$

$$\text{Creditor 2: } e - d_1 + \frac{1}{2}(d_1 + d_2 - e) = d_2 - \frac{d_1}{2} + \frac{1}{2}(e - d_2).$$

We saw previously that when $e = d_2$, each creditor has a loss of $\frac{d_1}{2}$ relative to
the debt she is owed. The part of the estate above d_2 is split equally, so the two
creditors' losses remain equal (Fig. 4.5).

We conclude that *the Contested Garment Rule linearly interpolates between Equal Division of Gains for $e \leq d_1$ (small estates) and Equal Division of Losses for $d_2 \leq e$ (large estates).*

4.4.4 More than Half is Like the Whole

Aumann and Maschler suggest that the Contested Garment Rule is perhaps related to the Talmudic principle that "more than half is like the whole, whereas less than half is like nothing." This principle says that the dividing line between two approaches to a problem is at the number one-half.

For example, in Talmudic law, a lender normally has an automatic lien on a borrower's property. However, in some cases, if the property is worth less than half the loan and the borrower is unable to repay it, the lender may not take the borrower's property (Arakhin 23b). Rashi explains that since the property is grossly inadequate to repay the loan, the lender has presumably relied not on the property but on the borrower's character for repayment, so the lender has no lien on the borrower's property.

In other words, if the property is worth less than half the loan and the borrower, despite his character, defaults, then the lender does not expect the loan to be repaid, so any repayment the lender receives is a gain relative to her expectation. If the property is worth more than half the loan, the lender expects the loan to be repaid, so any repayment she does not receive is a loss relative to her expectation.

The Contested Garment Rule can be seen as a sophisticated alternative to "more than half is like the whole." One principle is used to divide a small estate, and another is used to divide a large estate, but in between, one linearly interpolates between the two approaches.

(Arakhin 23b does not mean that the borrower is free of the obligation to repay the loan. Talmudic law does not have a concept of bankruptcy. If someone is unable to repay a loan, the lender cannot be forced to cancel it; it is the community's responsibility to help its destitute members, not the lender's. Should the borrower's circumstances improve, he must repay the lender.)

4.5 Kaminski's Proof of the Aumann–Maschler Theorem Using Glassware

Figure 4.6 is a schematic diagram of a piece of glassware. There are two glasses, of volumes d_1 and d_2. Each glass is equally divided between a top and a bottom connected by a small stem. In addition, a glass tube connects the bottoms of the two glasses. We will assume the stems and tube are very narrow, so that their volumes are negligible. The first glass represents the claim of Creditor 1, the second the claim of Creditor 2.

Fig. 4.6 Glassware for the
Contested Garment Rule

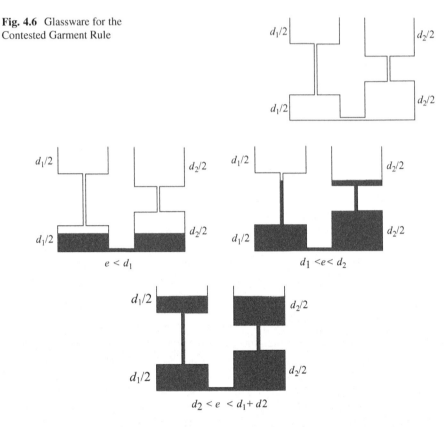

Fig. 4.7 Using the glassware to divide an estate with two creditors

Suppose one pours a volume e of liquid, $0 < e < d_1 + d_2$, into this glassware. The liquid represents the estate. Because of the connecting tube at the bottom, the liquid will rise to the same height in both glasses. If you ignore the liquid in the stems and the connecting tube, you will see that each creditor gets the amount to which she is entitled by the Contested Garment Rule. For $e \leq d_1$, the liquid divides equally. For $d_1 < e \leq d_2$, Creditor 1 gets $\frac{d_1}{2}$, and the rest goes to Creditor 2. For $d_2 < e < d_1 + d_2$, each glass fills to within the same distance of the top, so each creditor has an equal loss (Fig. 4.7).

The Aumann–Maschler Theorem can now be proved using more elaborate glassware. Given claims $d_1, \ldots, d_n$, construct the glassware shown in Fig. 4.8.

Pour in an amount e of liquid. It will rise to the same height in each glass. Since the height is the same in each pair of glasses, this division is consistent with the Contested Garment Rule. The division is unique: if we raise the height in one glass, in order to stay consistent with the Contested Garment Rule we must raise the height in all, so the total amount of liquid will increase.

Fig. 4.8 Glassware for an
estate with n creditors

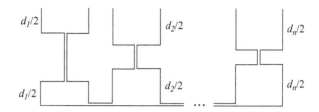

Kaminski learned about the Talmud's estate-division problem in a class taught
by the game theorist Peyton Young at the University of Maryland. Young was explaining his concept of "parametric representation" of allocation methods. Kaminski
writes [4]: "Sitting in class, I was repeatedly failing to visualize the parametric
representation of the Talmudic solution, and, displeased with myself, I stopped
listening and started thinking about an alternative. The 'hydraulic' idea came to my
mind in one of those unexplainable flashes. Later, I proved that in fact it is closely
related to parametric representation."

4.6 The Aumann–Maschler Theorem and Game Theory

A *cooperative game* consists of

1. A set of players $\{1,\ldots,n\}$
2. A value $V > 0$ to be divided among the players
3. A value function v from the power set of $\{1,\ldots,n\}$ into the nonnegative real
   numbers

In this context, a subset S of $\{1,\ldots,n\}$ is called a *coalition*. The number $v(S)$ is
interpreted as the part of the value V that the coalition S can get for itself no matter
what the other players do. Because of this interpretation, the value function v is
required to satisfy the following conditions.

1. $v(\emptyset) = 0$
2. $v(\{1,\ldots,n\}) = V$
3. If S_1 and S_2 are disjoint, then $v(S_1) + v(S_2) \leq v(S_1 \cup S_2)$

An *allocation* of the value V to the players is a vector $x = (x_1,\ldots,x_n)$ such that
all $x_i \geq 0$ and $x_1 + \cdots + x_n = V$. The problem of cooperative game theory is to choose
the allocation. There are various ideas about how to do it. We will only discuss one
of them..

Given an allocation x, the coalition S achieves the *excess* $e(x,S) = \sum_{j \in S} x_j - v(S)$.
Coalitions with low excess will presumably complain that they have been treated
unfairly and will not agree to the allocation. Perhaps one should choose x to avoid
small excesses as much as possible and thus minimize the complaining.

More precisely, given an allocation x, calculate all $2^n - 2$ excesses $e(x, S)$. (We ignore the empty set and the set $\{1, \ldots, n\}$.) Order them from smallest to largest to form an *excess vector* $e \in \mathbb{R}^{2^n - 2}$.

Given two excess vectors we can ask which precedes the other in the lexicographic ordering, which is defined as follows. Let $x = (x_1, \ldots, x_{2n-2})$ and $y = (y_1, \ldots, y_{2n-2})$ be two excess vectors, and suppose they first differ in the ith place. Then x precedes y in the lexicographic ordering if $x_i < y_i$.

- Example: $(1, 2, 4, 5)$ precedes $(2, 1, 2, 7)$
- Example: $(2, 2, 2, 7)$ precedes $(2, 2, 3, 6)$

Definition 4.6.1. The *nucleolus* of a cooperative game is the allocation whose excess vector is last in the lexicographic ordering.

Theorem 4.6.2. *Every cooperative game has a unique nucleolus.*

To find the nucleolus of a cooperative game, start with any allocation and adjust it to make one whose excess vector follows the excess vector of the first in the lexicographic ordering. When you can't go farther, you have found the nucleolus.

What does this have to do with estate-division problems? Associated with any estate division problem is a cooperative game. The value to be divided is the estate e. To define the value function, assume that any coalition can guarantee itself the larger of 0 and the amount that remains if all other creditors are paid in full.

Let's work this out for the second example in our Mishna. A man dies leaving an estate of 200. There are three creditors with claims of 100, 200, and 300. Any coalition can guarantee itself the larger of 0 and whatever is left after those not in the coalition are paid in full. Therefore

$$v(\{1\}) = 0, \ v(\{2\}) = 0, \ v(\{3\}) = 0, \ v(\{1,2\}) = 0, \ v(\{1,3\}) = 0, \ v(\{2,3\}) = 100.$$

We claim that the nucleolus of this game is the allocation proposed by the Mishna: $(50, 75, 75)$. To see this we consider the following table of excesses for an arbitrary allocation (x_1, x_2, x_3) and for the allocation proposed by the Mishna.

S	$v(S)$	$e((x_1, x_2, x_3), S)$	$e((50, 75, 75), S)$
$\{1\}$	0	x_1	50
$\{2\}$	0	x_2	75
$\{3\}$	0	x_3	75
$\{1,2\}$	0	$x_1 + x_2$	125
$\{1,3\}$	0	$x_1 + x_3$	125
$\{2,3\}$	100	$x_2 + x_3 - 100$	50

The excess vector is $(50, 50, 75, 75, 125, 125)$.

Can we adjust the allocation to make one whose excess vector follows this one in the lexicographic ordering? If we take anything from Creditor 1, the first 50 in the table will fall, so the new excess vector will *precede* the old one in the lexicographic ordering. If we take anything from Creditor 2 or 3 and give it to Creditor 1, the other 50 in the table will fall, so the new excess vector will again precede the old one in the lexicographic ordering. The only remaining possibility is to take something from Creditor 2 or 3 and give it to the other. If we do this, the two 50s in the table will remain, but one of the 75s will decrease. Again the new excess vector will precede the old one in the lexicographic ordering. We conclude that the allocation $(50, 75, 75)$ proposed by the Mishna is the nucleolus of the associated cooperative game.

Theorem 4.6.3 (Aumann–Maschler). *In any estate-division problem, the unique allocation that is consistent with the Contested Garment Rule is also the nucleolus of the associated cooperative game.*

It is a remarkable fact that Aumann and Maschler discovered the relationship between the Talmud's proposed allocations and the nucleolus before they thought of the relation to the Contested Garment Rule. Here is the story as told by Aumann in [2]: "Mike and I sat down to try to figure out what is going on in that passage. We put the nine relevant numbers on the blackboard in tabular form and gazed at them mutely. There seemed no rhyme or reason to them—not equal, not proportional, nothing. We tried the Shapley value of the corresponding coalitional game; this, too, did not work. Finally one of us said, let's try the nucleolus; to which the other responded, come on, that's crazy, the nucleolus is an extremely sophisticated notion of modern mathematical game theory, there's no way that the sages of the Talmud could possibly have thought of it. What do you care, said the first; it will cost us just 15 min of calculation. So we did the calculation, and the nine numbers came out precisely as in the Talmud!" They then discovered by a literature search that the nucleolus had recently been proved to have a consistency property: if you look at the amounts assigned by the nucleolus to a subset of players, this is precisely the nucleolus of the reduced game with only those players and value equal to the total assigned to them by the nucleolus of the original game.

4.7 Final Remark

There is one aspect of the Aumann–Maschler solution that bothered me. Their solution has nothing whatever to do with proportional division. Nevertheless, among the three examples of estate division given in the Talmud, one, the last, is proportional: each creditor gets exactly half her claim. Was the Talmud trying to lead us astray?

Joseph Bak of the City College of New York saw the slides for a talk I had given in which I asked this question. He sent the following answer: in any estate-division problem, if the estate is exactly half the total of the debts, then the Aumann–Maschler rule will produce proportional division. In fact each debtor will get exactly half what she is owed.

This is easy to see from Fig. 4.8. The volumes of the bottoms of all the glasses add up to $\frac{1}{2}(d_1 + \cdots + d_n)$. If this is the amount of the estate, it will exactly fill all the bottoms. Thus the ith debtor gets $\frac{d_i}{2}$.

If one gives examples of the Aumann–Maschler estate-division rule using nice round numbers like 100, 200, etc. for the estate and the debts, it is quite easy for one of the examples to have the estate equal to exactly half the total debts, thus producing proportional division. This may be what happened in the Talmud.

References

1. Aumann, R.J., Maschler, M.: Game theoretic analysis of a bankruptcy problem from the Talmud. J. Econ. Theor. **36**, 195–213 (1985)
2. Aumann, R.J.: Working with Mike, in Michael Maschler: In Memoriam. Center for the Study of Rationality, Hebrew University of Jerusalem (2008)
3. Kaminski, M.: Hydraulic rationing. Math. Soc. Sci. **40**, 131–155 (2000)
4. Kaminski, M.: personal communication.

Chapter 5
On the Non-uniqueness of the Decomposition of Weighted Pseudo Almost Periodic Functions

Gaston M. N'Guérékata

Abstract In this short communication, we show through an example that the decomposition of a ρ-weighted pseudo almost periodic function is not unique when $\inf_{t \in \mathbb{R}} \rho(t) = 0$.

Keywords Weighted pseudo almost periodicity

Mathematics Subject Classification (1991): 34G10, 47D06

5.1 Introduction

The concept of weighted pseudo almost periodicity was introduced by Diagana in [4, 7]. It is a generalization of Zhang's pseudo almost periodicity [11, 12, 14, 15]. Since then, the concept has attracted several authors especially in connection with the existence of solutions of differential equations [1, 2, 5–7, 13]. There is a recent generalization to almost automorphic case by Blot et al. [3]. See also [9, 10] for more recent developments.

In this paper, we like to discuss the uniqueness of the decomposition of weighted pseudo almost periodic functions as presented in [1] and many others. Note that this problem has been addressed in [8].

G.M. N'Guérékata (✉)
Department of Mathematics, Morgan State University, 1700 E. Cold Spring Lane, Baltimore, MD 21251, USA
e-mail: Gaston.N'Guerekata@morgan.edu

B. Toni et al. (eds.), *Bridging Mathematics, Statistics, Engineering and Technology*, Springer Proceedings in Mathematics & Statistics 24, DOI 10.1007/978-1-4614-4559-3_5, © Springer Science+Business Media New York 2012

5.2   The Result

Like in [5], let $\mathbb{U}$ be the collection of functions (weights) $\rho : \mathbb{R} \to (0, \infty)$, which are locally integrable over $\mathbb{R}$ such that $\rho(x) > 0$ almost everywhere. Set

$$m(T, \rho) := \int_{-T}^{T} \rho(t)dt,$$

$$\mathbb{U}_\infty := \left\{ \rho \in \mathbb{U} : \lim_{T \to \infty} m(T, \rho) = \infty \right\},$$

$$\mathbb{U}_B := \left\{ \rho \in \mathbb{U}_\infty : \rho \text{ is bounded with } \inf_{t \in \mathbb{R}} \rho(t) > 0 \right\}.$$

Obviously, $\mathbb{U}_B \subset \mathbb{U}_\infty \subset \mathbb{U}$, with strict inclusions.

For each $\rho \in \mathbb{U}_\infty$, define

$$PAP_0(X, \rho) := \left\{ f \in BC(\mathbb{R}, X) : \lim_{T \to \infty} \frac{1}{m(T, \rho)} \int_{-T}^{T} \|f(t)\| \rho(t)dt = 0 \right\}.$$

Definition 5.2.1. Let $\rho \in \mathbb{U}_\infty$. A function $f \in BC(\mathbb{R}, X)$ is called weighted pseudo almost periodic or ρ-pseudo almost periodic if it can be expressed as $f = g + \varphi$, where $g \in AP(X)$ and $\varphi \in PAP_0(X, \rho)$. The collection of such functions will be denoted by $PAP(X, \rho)$.

Theorem 5.2.2 ([5] Theorem 3.1). *Fix $\rho \in \mathbb{U}_\infty$. The decomposition of a ρ-pseudo almost periodic function $f = g + \phi$, $g \in AP(X)$, $\phi \in PAP_0(X, \rho)$ is unique.*

We would like to revisit this result when $\inf_{t \in \mathbb{R}} \rho(t) = 0$. Indeed, let $X = \mathbb{R}$ and for each $n \in \mathbb{Z}$,

$$\rho(t) = \begin{cases} 1, & t \in [n, n + \frac{1}{2}], \\ e^{-t^2}, & t \in (n + \frac{1}{2}, n + 1). \end{cases}$$

Clearly $\inf_{t \in \mathbb{R}} \rho(t) = 0$.

Since $\rho(t) = 1$ on each $[n, n + \frac{1}{2}]$, it follows that

$$m(T, \rho) = \int_{-T}^{T} \rho(t)dt \geq \int_{-[T]}^{[T]} \rho(t)dt$$

$$= \sum_{k=1}^{2[T]} \int_{-[T]+k-1}^{-[T]+k} \rho(t)dt$$

$$\geq \sum_{k=1}^{2[T]} \int_{-[T]+k-1}^{-[T]+k-\frac{1}{2}} \rho(t)dt$$

$$\geq \sum_{k=1}^{2[T]} \frac{1}{2} = [T].$$

Thus $\rho \in \mathbb{U}_\infty$.

Define x periodically on $\mathbb{R}$ (with period 1) from

$$x(t) = \begin{cases} 0, & t \in [0, \frac{1}{2}], \\ 4t - 2, & t \in [\frac{1}{2}, \frac{3}{4}], \\ -4t + 4, & t \in [\frac{3}{4}, 1], \end{cases}$$

Thus $x(t+1) = x(t)$ for all $t \in \mathbb{R}$. Obviously, x is continuous. So $x \in AP(\mathbb{R})$.

On the other hand, let $T > 0$ and choose $n > T$. Then using the fact that $x(t) = 0$ on each $[n, n + \frac{1}{2}]$ we have

$$0 \leq \int_{-T}^{T} x(t)\rho(t)dt \leq \int_{-n}^{n} x(t)\rho(t)dt$$

$$= \sum_{k=0}^{k=2n-1} \int_{-n+k+\frac{1}{2}}^{-n+k+1} e^{-t^2}dt$$

$$\leq \int_{-n}^{n} e^{-t^2}dt \leq \sqrt{\pi}.$$

Then,

$$\lim_{T \to \infty} \frac{\int_{-T}^{T} x(t)\rho(t)dt}{\int_{-T}^{T} \rho(t)dt} = 0,$$

which means that $x \in PAP_0(\mathbb{R}, \rho)$.

Now, we get $x \in AP(\mathbb{R}) \cap PAP_0(\mathbb{R}, \rho)$. Obviously, $x \in PAP(\mathbb{R}, \rho)$. However, the decomposition is not unique since

$$x(t) = x(t) + 0 = 0 + x(t) = 2x(t) - x(t) = \cdots.$$

From the above we get:

Theorem 5.2.3. *The set*

$$AP(X) \cap PAP_0(X, \rho)$$

is nontrivial if $\rho \in \mathbb{U}_\infty$.

Remark 5.2.4. Note that in our example $\inf_{t \in \mathbb{R}} \rho(t) = 0$. *Thus* $\rho(t) \notin \mathbb{U}_B$. *It has been shown in [8] that uniqueness of the decomposition occurs when* $PAP_0(X, \rho)$ *is translation invariant. The problem remains open if this is not true.*

References

1. Agarwal, R.P., Diagana, T., Hernández, M.E.: Weighted pseudo almost periodic solutions to some partial neutral functional differential equations. J. Nonlinear and Convex Analysis **8** (3), 397–415 (2007)
2. Agarwal, R.P., de Andrade, B., Cuevas, C.: Weighted pseudo almost periodic solutions of a class of semilinear fractional differential equations. Nonlinear Analysis: Real Word Applications (in press).
3. Blot, J., Mophou, G.M., N'Guérékata, G.M., Pennequin, D.: Weighted pseudo almost automorphic functions and applications to abstract differential equations. Nonlinear Anal. **71** 3–4, 903–909 (2009)
4. Diagana, T.: Weighted pseudo almost periodic functions and applications. C. R. Acad. Sci. Paris, Ser.I **343**(10), 643–646 (2006)
5. Diagana, T.: Weighted pseudo almost periodic solutions to some differential equations. Nonlinear Anal. **68**, 2250–2260 (2008)
6. Diagana, T.: Existence of weighted pseudo almost periodic solutions to some non-autonomous differential equations. Intern. J. Evol. Equ. **2**(4), 397–410 (2008)
7. Diagana, T.: Weighted pseudo almost periodic solutions to a neutral delay integral equation of advanced type. Nonlinear Anal. **70,** 298–304 (2009)
8. Liang, J., Xiao, T-J., Zhang, J.: Decomposition of weighted pseudo almost periodic functions. Nonlinear Anal. **73**, 3456–3461 (2010)
9. Liu, J.-H., Song, X.-Q.: Almost automorphic and weighted pseudo almost automorphic solutions of semilinear evolutions equations. J. Funct. Anal. **258**, 196–207 (2010)
10. Liu, J.H., N'Guérékata, G.M., Van Minh, N.: Topics on stability and periodicity in abstract differential equations, Series on Concrete and Applicable Mathemaics. World Scientific, New jersey-London-Singapore (2008)
11. Mophou, G.: Weighted pseudo almost automorphic mild solutions to semilinear fractional differential equations. Appl. Math. Comput. **217**, 7579–7587 (2011)
12. N'Guérékata, G.M.: Almost Periodic and Almost Automorphic Functions in Abstract Spaces. Kluwer Academic/Plenum Publishers, New York-London-Moscow-Dordrecht (2001)
13. Zhang, C.: Pseudo almost periodic functions and their applications. Ph. D. Thesis, University of Western Ontarion (1992)
14. Zhang, C.: Pseudo almost periodic solutions of some differential equations. J. Math. Anal. Appl. **181**, 62–76 (1994)
15. Zhang L., Xu, Y.: Weighted pseudo almsot periodic solutions for functional differential equations. Elec. J. Diff. Equ. **2007**(146), 1–7 (2007)

Chapter 6
Note on the Almost Periodic Stochastic Beverton–Holt Equation

Paul H. Bezandry

Abstract In this paper almost periodic random sequence is defined and investigated. It is then applied to study the existence and uniqueness of the almost periodic solution of the stochastic Beverton–Holt equation with varying survival rates and intrinsic growth rates.

Keywords Almost periodicity in mean • Difference equation • Beverton-Holt equation

Mathematics Subject Classification (2000): Primary: 60H05, 60H15; Secondary: 34G20, 43A60

6.1 Introduction

In constant environments, theoretical discrete-time population models are usually formulated under the assumption that the dynamics of the total population size in generation n, denoted by $X(n)$, are governed by equations of the form

$$X(n+1) = \gamma X(n) + f(X(n)), \tag{6.1}$$

where $\gamma \in (0,1)$ is the constant "probability" of surviving per generation, and $f : \mathbb{R}_+ \to \mathbb{R}_+$ models the recruitment process.

Almost periodic effects can be introduced into Eq. (6.1) by writing the recruitment function or the survival probability as almost periodic sequences. This is model with the equation

P.H. Bezandry (✉)
Department of Mathematics, Howard University, Washington, DC 20059, USA
e-mail: pbezandry@howard.edu

B. Toni et al. (eds.), *Bridging Mathematics, Statistics, Engineering and Technology,*
Springer Proceedings in Mathematics & Statistics 24, DOI 10.1007/978-1-4614-4559-3_6,
© Springer Science+Business Media New York 2012

$$X(n+1) = \gamma_n X(n) + f(n, X(n)), \tag{6.2}$$

where either $\{\gamma_n\}_{n \in \mathbb{Z}_+}$ or $n \to f(n, X(n))$ is almost periodic and each $\gamma_n \in (0,1)$.

In their paper, Franke and Yakubu [4] studied (6.2) with the periodic Beverton–Holt recruitment function

$$f(n, X(n)) = \frac{(1 - \gamma_n)\mu K_n X(n)}{(1 - \gamma_n)K_n + (\mu - 1 + \gamma_n)X(n)}, \tag{6.3}$$

where the carrying capacity K_n is p-periodic, $K_{n+p} = K_n$ for all $n \in \mathbb{Z}_+$, and $\mu > 1$.

In this paper, we assume that the intrinsic growth rate μ, the carrying capacity K, and the survival rate γ vary and that K and γ are random. Equation (6.3) becomes

$$X(n+1) = \gamma_n X(n) + \frac{(1 - \gamma_n)\mu_n K_n X(n)}{(1 - \gamma_n)K_n + (\mu_n - 1 + \gamma_n)X(n)}. \tag{6.4}$$

We are then concerned with the existence of almost periodic solutions to Eq. (6.4). The case where the intrinsic growth rate is constant has been treated in [2]. The assumption of randomness makes a lot of sense in this context since nature is basically stochastic rather than deterministic.

The paper is organized as follows. In Sect. 6.2, we recall a basic theory of almost periodic random sequences on $\mathbb{Z}_+$. In Sect. 6.3, we apply the techniques developed in Sect. 6.2 to find some sufficient conditions for the existence and uniqueness of the almost periodic solution to the stochastic Beverton–Holt difference equation with varying survival rates and intrinsic growth rates.

6.2 Preliminaries

In this section we establish a basic theory for almost periodic random sequences. To facilitate our task, we first introduce the notations needed in the sequel.

Let $(\mathbb{B}, \|\cdot\|)$ be a Banach space and let $(\Omega, \mathcal{F}, \mathbf{P})$ be a complete probability space. Throughout the rest of the paper, $\mathbb{Z}_+$ denotes the set of all nonnegative integers. Define $L^1(\Omega; \mathbb{B})$ to be the space of all $\mathbb{B}$-valued random variables V such that

$$\mathbf{E}\|V\| := \left(\int_\Omega \|V(\omega)\| d\mathbf{P}(\omega) \right) < \infty. \tag{6.5}$$

It is then routine to check that $L^1(\Omega; \mathbb{B})$ is a Banach space when it is equipped with its natural norm $\|\cdot\|_1$ defined by $\|V\|_1 := \mathbf{E}\|V\|$ for each $V \in L^1(\Omega, \mathbb{B})$.

Let $X = \{X_n\}_{n \in \mathbb{Z}_+}$ be a sequence of $\mathbb{B}$-valued random variables satisfying $\mathbf{E}\|X_n\| < \infty$ for each $n \in \mathbb{Z}_+$. Thus, interchangeably we can, and do, speak of such a sequence as a function, which goes from $\mathbb{Z}_+$ into $L^1(\Omega; \mathbb{B})$.

This setting requires the following preliminary definitions.

Definition 6.2.1. A $\mathbb{B}$-valued sequence $x = \{x(n)\}_{n \in \mathbb{Z}_+}$ is said to be Bohr almost periodic if for each $\varepsilon > 0$ there exists $N_0(\varepsilon) > 0$ such that among any N_0 consecutive integers there exists at least an integer $p > 0$ for which

$$\|X(n+p) - X(n)\| < \varepsilon, \quad \forall n \in \mathbb{Z}_+.$$

Definition 6.2.2. An $L^1(\Omega; \mathbb{B})$-valued random sequence $X = \{X(n)\}_{n \in \mathbb{Z}_+}$ is said to be Bohr almost periodic in mean if for each $\varepsilon > 0$ there exists $N_0(\varepsilon) > 0$ such that among any N_0 consecutive integers there exists at least an integer $p > 0$ for which

$$\mathbf{E}\|X(n+p) - X(n)\| < \varepsilon, \quad \forall n \in \mathbb{Z}_+.$$

An integer $p > 0$ with the above-mentioned property is called an ε-almost period for X. The collection of all $\mathbb{B}$-valued random sequences $X = \{X(n)\}_{n \in \mathbb{Z}_+}$ which are Bohr almost periodic in mean is then denoted by $\mathrm{AP}(\mathbb{Z}_+; L^1(\Omega; \mathbb{B}))$.

Similarly, one defines the Bochner almost periodicity in mean as follows.

Definition 6.2.3. An $L^1(\Omega; \mathbb{B})$-valued random sequence $X = \{X(n)\}_{n \in \mathbb{Z}_+}$ is called mean Bochner almost periodic if for every sequence $\{m_k\}_{k \in \mathbb{Z}_+} \subset \mathbb{Z}_+$ there exists a subsequence $\{m'_k\}_{k \in \mathbb{Z}_+}$ such that $\{X(n+m'_k))\}_{k \in \mathbb{Z}_+}$ converges (in the mean) uniformly in $n \in \mathbb{Z}_+$.

Following along the same arguments as in the proof of [3, Theorem 2.4, p 241], one can show that those two notions of almost periodicity coincide.

Theorem 6.2.4. *An $L^1(\Omega; \mathbb{B})$-valued random sequence $X = \{X(n)\}_{n \in \mathbb{Z}_+}$ is Bochner almost periodic in mean if and only if it is Bohr almost periodic in mean.*

An important and straightforward consequence of Theorem 6.2.4 is the next corollary, which pays a key role in the proof of Theorem 6.3.1.

Corollary 6.2.5 ([1]). *If $X_1 = \{X^1(n)\}_{n \in \mathbb{Z}_+}, X_2 = \{X^1(n)\}_{n \in \mathbb{Z}_+}, \ldots,$ and $X_N = \{X^N(n)\}_{n \in \mathbb{Z}_+}$ are N random sequences, which belong to $\mathrm{AP}(\mathbb{Z}_+; L^1(\Omega, \mathbb{B}))$, then for each $\varepsilon > 0$ there exists $N_0(\varepsilon) > 0$ such that among any $N_0(\varepsilon)$ consecutive integers there exists an integer $p > 0$ for which*

$$\mathbf{E}\|X^j(n+p) - X(n)\| < \varepsilon$$

for $n \in \mathbb{Z}_+$ and for $j = 1, 2, \ldots, N$.

Lemma 6.2.6 ([1]). *If X belongs to $\mathrm{AP}(\mathbb{Z}_+; L^1(\Omega; \mathbb{B}))$, then there exists a constant $M > 0$ such that $\mathbf{E}\|X(n)\| \le M$ for each $n \in \mathbb{Z}_+$.*

Lemma 6.2.7 ([1]). *Let $\mathbb{B} = \mathbb{R}$. If the sequences X and Y are (stochastically) independent one another and both belong to $\mathrm{AP}(\mathbb{Z}_+; L^1(\Omega; \mathbb{R}))$, then the sequence $XY = \{X(n)Y(n), n \in \mathbb{Z}_+\}$ belongs to $\mathrm{AP}(\mathbb{Z}_+; L^1(\Omega; \mathbb{R}))$.*

Let $(\mathbb{B}_1, \|\cdot\|_1)$ and $(\mathbb{B}_2, \|\cdot\|_2)$ be Banach spaces and let $L^1(\Omega; \mathbb{B}_1)$ and $L^1(\Omega; \mathbb{B}_2)$ be their corresponding L^1-spaces, respectively.

Definition 6.2.8. A function $F : \mathbb{Z}_+ \times L^1(\Omega; \mathbb{B}_1) \mapsto L^1(\Omega; \mathbb{B}_2), (n, U) \mapsto F(n, U)$ is said to be almost periodic in mean in $n \in \mathbb{Z}_+$ uniformly in $U \in \mathbb{K}$ where

$\mathbb{K} \subset L^1(\Omega; \mathbb{B}_1)$ is compact if for any $\varepsilon > 0$, there exists a positive integer $l(\varepsilon, \mathbb{K})$ such that among any l consecutive integers there exists at least a integer p with the following property:

$$\mathbf{E}\|F(n+p,U) - F(n,U)\|_2 < \varepsilon$$

for each random variable $U \in \mathbb{K}$ and $n \in \mathbb{Z}^+$.

Here again, the number p will be called an ε-translation of F and the set of all ε-translations of F is denoted by $\mathcal{E}(\varepsilon, F, \mathbb{K})$.

Let $\mathrm{UB}(\mathbb{Z}_+; L^1(\Omega; \mathbb{B}))$ denote the collection of all uniformly bounded $L^1(\Omega; \mathbb{B})$-valued random sequences $X = \{X(n)\}_{n \in \mathbb{Z}_+}$. It is then easy to check that the space $\mathrm{UB}(\mathbb{Z}_+; L^1(\Omega; \mathbb{B}))$ is a Banach space when it is equipped with the norm:

$$\|X\|_\infty = \sup_{n \in \mathbb{Z}_+} \mathbf{E}\|X(n)\|.$$

Lemma 6.2.9 ([1]). $\mathrm{AP}(\mathbb{Z}_+; L^1(\Omega; \mathbb{B})) \subset \mathrm{UB}(\mathbb{Z}_+; L^1(\Omega; \mathbb{B}))$ *is a closed space.*

In view of the above, the space $\mathrm{AP}(\mathbb{Z}_+; L^1(\Omega; \mathbb{B}))$ of almost periodic random sequences equipped with the sup norm $\|\cdot\|_\infty$ is also a Banach space.

We now state the following composition result.

Theorem 6.2.10 ([1]). *Let* $F : \mathbb{Z}_+ \times L^1(\Omega; \mathbb{B}_1) \mapsto L^1(\Omega; \mathbb{B}_2)$, $(n, U) \mapsto F(n, U)$ *be almost periodic in mean in* $n \in \mathbb{Z}_+$ *uniformly in* $U \in L^1(\Omega; \mathbb{B}_1)$. *If in addition, F is Lipschitz in* $U \in \mathbb{K}$, *where* $\mathbb{K} \subset L^1(\Omega; \mathbb{B}_1)$ *is compact, that is, there exists* $L > 0$ *such that*

$$\mathbf{E}\|F(t,U) - F(t,V)\|_2 \leq M \; \mathbf{E}\|U - V\|_1 \; \forall U, V \in L^1(\Omega; \mathbb{B}_1), n \in \mathbb{Z}_+)$$

then for any almost periodic random sequence $X = \{X(n)\}_{n \in \mathbb{Z}_+}$, *then the* $L^1(\Omega; \mathbb{B}_1)$-*valued random sequence* $Y(n) = F(n, X(n))$ *is almost periodic in mean.*

6.3 Application to Stochastic Beverton–Holt Equation

In this section, we assume that both carrying capacity K_n and the survival rate γ_n are random and that $\{\gamma_n, n \in \mathbb{Z}_+\}$ are (stochastically) independent and (stochastically) independent of the sequence K_n, $n \in \mathbb{Z}_+$.

We have the following theorem.

Theorem 6.3.1. *Suppose that* $\{\mu_n\}_{n \in \mathbb{Z}_+}$ *is almost periodic and that both* $\{\gamma_n\}_{n \in \mathbb{Z}_+}$ *and* $\{K_n\}_{n \in \mathbb{Z}_+}$ *are almost periodic in mean. Then Eq (6.4) has a unique mean almost periodic solution whenever*

$$\sup_{n \in \mathbb{Z}_+} \mathbf{E}[\gamma_n] < \frac{1}{\mu + 1},$$

where $\mu := \sup_{n \in \mathbb{Z}_+} \mu_n$.

Remark 6.3.2. In the above theorem, μ is well defined since $\{\mu_n\}_{n\in\mathbb{Z}_+}$ is almost periodic.

The proof of Theorem 6.3.1 requires the following lemma.

Lemma 6.3.3. *Let*

$$f(n,X(n)) = \frac{(1-\gamma_n)\mu_n K_n X(n)}{(1-\gamma_n)K_n + (\mu_n - 1 + \gamma_n)X(n)}$$

where $\{\mu_n\}_{n\in\mathbb{Z}_+}$ is almost periodic and both $\{K_n\}_{n\in\mathbb{Z}_+}$ and $\{\gamma_n\}_{n\in\mathbb{Z}_+}$ are almost periodic in mean, and let μ be as in Theorem 6.3.1. *Then,*

(i) f is μ-Lipschitz in the following sense:

$$\mathbf{E}|f(n,U) - f(n,V)| \leq \mu \ \mathbf{E}|U - V|, \ \forall U,V \in L^1(\Omega;\mathbb{R}_+), n \in \mathbb{Z}_+;$$

(ii) If X belongs to $AP(\mathbb{Z}_+;L^1(\Omega;\mathbb{R}_+))$, then the sequence $\{f(n,X(n))\}_{n\in\mathbb{Z}_+}$ also belongs to $AP(\mathbb{Z}_+;L^1(\Omega;\mathbb{R}_+))$.

Proof. (Lemma 6.3.3)

It is a routine to show that

$$|f(n,U) - f(n,V)| \leq \mu_n|U - V|,$$

and hence

$$\mathbf{E}|f(n,U) - f(n,V)| \leq \mu \ \mathbf{E}|U - V|.$$

To prove the almost periodicity of $n \rightarrow f(n,X(n))$, set $A_n = (1 - \gamma_n)K_n$ and $B_n = \mu_n - 1 + \gamma_n$. Then f can be written as follows:

$$f(n,X(n)) = \mu_n \frac{A_n X(n)}{A_n + B_n X(n)} \quad \text{for each } n \in \mathbb{Z}_+.$$

Using the fact that $\{\mu_n\}$ is almost periodic and that $\{\gamma_n\}$ and $\{K_n\}$ are almost periodic in mean, and making use of Lemma 6.2.6 and Corollary 6.2.5, we can choose a constant $K > 0$ such that $\mathbf{E}|K_n| < K$ for all $n \in \mathbb{Z}_+$ and for each $\varepsilon > 0$ there exists a positive integer $N_0(\varepsilon)$ such that among any $N_0(\varepsilon)$ consecutive integers, there exists an integer $p > 0$, a common ε-almost period for $\{\mu_n\}$, $\{\gamma_n\}$, and $\{K_n\}$ for which

$$|\mu_{n+p} - \mu_n| \leq \frac{\varepsilon}{3\mu M_1} \quad \mathbf{E}|\gamma_{n+p} - \gamma_n| \leq \frac{\varepsilon}{3\mu M_2}, \text{and} \ \mathbf{E}|K_{n+p} - K_n| \leq \frac{\varepsilon}{3\mu M_3},$$

where $M_1 = \frac{K}{(m-1)^2}$, $M_2 = K\left[\frac{1}{m-1} + \frac{1}{(m-1)^2}\right]$, and $M_3 = \frac{2}{m-1} + \frac{\mu}{(m-1)^2}$.

Observe that

$$\mathbf{E}|f(n+p,U) - f(n,U)| = \mu_n \ \mathbf{E} \left| \frac{A_{n+p}U}{A_{n+p}+B_{n+p}U} - \frac{A_nU}{A_n+B_nU} \right|$$

$$\leq \mu \ \mathbf{E} \left| \frac{(A_{n+p}B_n - A_nB_{n+p})U^2}{B_{n+p}B_nU^2} \right| = \mu \ \mathbf{E} \left| \frac{A_{n+p}}{B_{n+p}} - \frac{A_n}{B_n} \right|.$$

We now evaluate $\mathbf{E} \left| \frac{A_{n+p}}{B_{n+p}} - \frac{A_n}{B_n} \right|$. Using the hypothesis of independence of the random sequences $\{\gamma_n\}_{n\in\mathbb{Z}_+}$ and $\{K_n\}_{n\in\mathbb{Z}_+}$, we have

$$\mathbf{E} \left| \frac{A_{n+p}}{B_{n+p}} - \frac{A_n}{B_n} \right| = \mathbf{E} \left| \frac{(1-\gamma_{n+p})K_{n+p}}{\mu_n - 1 + \gamma_{n+p}} - \frac{(1-\gamma_n)K_n}{\mu_n - 1 + \gamma_n} \right|$$

$$= \mathbf{E} \left| \frac{1}{(\mu_{n+p}-1+\gamma_{n+p})(\mu_n-1+\gamma_n)} \left[(\mu_n-1)(1-\gamma_{n+p})K_{n+p} \right. \right.$$

$$\left. \left. + \gamma_n(1-\gamma_{n+p})K_{n+p} - (\mu_{n+p}-1)(1-\gamma_n)K_n - \gamma_{n+p}(1-\gamma_n)K_n \right] \right|$$

$$= \mathbf{E} \left| \frac{1}{(\mu_{n+p}-1+\gamma_{n+p})(\mu_n-1+\gamma_n)} \left[(\mu_n-1)[K_{n+p}-K_n] \right. \right.$$

$$- [\mu_{n+p}-\mu_n]K_n + (\mu_n-1)[\gamma_{n+p}K_{n+p}-\gamma_nK_n] + \gamma_nk_n[\mu_{n+p}-\mu_n]$$

$$\left. \left. + \gamma_nK_{n+p} - \gamma_{n+p}K_n - \gamma_n\gamma_{n+p}[K_{n+p} - K_n] \right] \right|$$

$$= \mathbf{E} \left| \frac{\mu_n - 1}{(\mu_{n+p}-1+\gamma_{n+p})(\mu_n-1+\gamma_n)} [K_{n+p} - K_n] \right.$$

$$- \frac{K_n}{(\mu_{n+p}-1+\gamma_{n+p})(\mu_n-1+\gamma_n)} [\mu_{n+p} - \mu_n]$$

$$+ \frac{(\mu_n-1)K_n}{(\mu_{n+p}-1+\gamma_{n+p})(\mu_n-1+\gamma_n)} [\gamma_{n+p} - \gamma_n]$$

$$+ \frac{\mu_n}{(\mu_{n+p}-1+\gamma_{n+p})(\mu_n-1+\gamma_n)} [K_{n+p} - K_n]$$

$$+ \frac{\gamma_n}{(\mu_{n+p}-1+\gamma_{n+p})(\mu_n-1+\gamma_n)} [K_{n+p} - K_n]$$

$$\left. - \frac{K_n}{(\mu_{n+p}-1+\gamma_{n+p})(\mu_n-1+\gamma_n)} [\gamma_{n+p} - \gamma_n] \right|$$

$$\leq \frac{1}{\mu_{n+p} - 1} \ \mathbf{E}|K_{n+p}-K_n| + \frac{1}{(\mu_{n+p} - 1)(\mu_n - 1)} \ |\mu_{n+p}-\mu_n| \ \mathbf{E}[K_n]$$

$$+ \frac{1}{\mu_{n+p} - 1} \; \mathbf{E}[K_n] \; \mathbf{E}|\gamma_{n+p} - \gamma_n| + \frac{1}{\mu_{n+p} - 1} \; \mathbf{E}[\gamma_n] \; \mathbf{E}|K_{n+p} - K_n|.$$

$$+ \frac{1}{(\mu_{n+p} - 1)(\mu_n - 1)} \; \mathbf{E}[K_n] \cdot \mathbf{E}|\gamma_{n+p} - \gamma_n|$$

$$+ \frac{\mu_n}{(\mu_{n+p} - 1)(\mu_n - 1)} \; \mathbf{E}|K_{n+p} - K_n|$$

But the sequence $\{\mu_n\}$ is bounded. That is, there exist $m > 1$ and $\mu := \sup\limits_{n \in \mathbb{Z}_+} \mu_n$ such that $m \leq \mu_n \leq \mu$ for all $n \in \mathbb{Z}_+$. Thus, $\dfrac{1}{\mu_n - 1} \leq \dfrac{1}{m - 1}$ for all $n \in \mathbb{Z}_+$. Hence,

$$\mathbf{E}\left|\frac{A_{n+p}}{B_{n+p}} - \frac{A_n}{B_n}\right| \leq \frac{1}{m-1} \; \mathbf{E}|K_{n+p} - K_n| + \frac{K}{(m-1)^2} \; |\mu_{n+p} - \mu_n|$$

$$+ \frac{K}{m-1} \; \mathbf{E}|\gamma_{n+p} - \gamma_n| + \frac{1}{m-1} \; \mathbf{E}|K_{n+p} - K_n|.$$

$$+ \frac{K}{(m-1)^2} \; \mathbf{E}|\gamma_{n+p} - \gamma_n| + \frac{\mu}{(m-1)^2} \; \mathbf{E}|K_{n+p} - K_n|$$

$$\leq \left[\frac{2}{m-1} + \frac{\mu}{(m-1)^2}\right] \mathbf{E}|K_{n+p} - K_n| + \frac{K}{(m-1)^2} \; |\mu_{n+p} - \mu_n|$$

$$+ K \left[\frac{1}{m-1} + \frac{1}{(m-1)^2}\right] \mathbf{E}|\gamma_{n+p} - \gamma_n|$$

Thus, we obtain

$$\mathbf{E}|f(n+p, U) - f(n, U)| \leq \frac{\varepsilon}{3} + \frac{\varepsilon}{3} + \frac{\varepsilon}{3} = \varepsilon.$$

By Theorem 6.2.10, we can conclude that $n \to f(n, X(n))$ is almost periodic in mean. □

We now prove Theorem 6.3.1.

Proof. By Lemma 6.3.3(ii), if $u \in \mathrm{AP}(\mathbb{Z}_+, L^1(\Omega; \mathbb{R}_+))$, then $n \to f(n, u(n))$ belongs to $\mathrm{AP}(\mathbb{Z}_+, L^1(\Omega; \mathbb{R}_+))$. Define the nonlinear operator Γ by setting:

$$\Gamma : \mathrm{AP}(\mathbb{Z}_+, L^1(\Omega; \mathbb{R}_+)) \mapsto \mathrm{AP}(\mathbb{Z}_+, L^1(\Omega; \mathbb{R}_+)),$$

where

$$\Gamma u(n) := \sum_{r=0}^{n-1} \left(\prod_{s=r}^{n-1} \gamma_s \right) f(r, u(r)),$$

is the representation of the solution of Eq. (6.4).

It is clear that Γ is well defined. Now, let $u, v \in \text{AP}(\mathbb{Z}_+, L^1(\Omega; \mathbb{R}_+))$ having the same property as X defined in the Beverton–Holt equation. One can easily see that

$$\mathbf{E}|\Gamma u(n) - \Gamma v(n)| \leq \sum_{r=0}^{n-1} \left\{ \left(\prod_{s=r}^{n-1} \mathbf{E}|\gamma_s| \right) \mathbf{E}|f(r, u(r)) - f(r, v(r))| \right\},$$

and hence letting $\beta = \sup_{n \in \mathbb{Z}_+} \mathbf{E}[\gamma_n]$ we obtain

$$\sup_{n \in \mathbb{Z}_+} \mathbf{E}|\Gamma u(n) - \Gamma v(n)| \leq \left(\frac{\mu \beta}{1 - \beta} \right) \sup_{n \in \mathbb{Z}_+} \mathbf{E}|u(n) - v(n)|.$$

Obviously, Γ is a contraction whenever $\dfrac{\mu \beta}{1 - \beta} < 1$. In that event, using the Banach fixed point theorem it easily follows that Γ has a unique fixed point, $\overline{X}$, which obviously is the unique mean almost periodic solution of (6.4). $\qquad \square$

References

1. Bezandry, P., Diagana, T.: Almost Periodic Stochastic Processes. Springer, New York (2011)
2. Bezandry, P., Diagana, T., Elaydi, S.: On the stochastic Beverton-Holt equation with survival rates. J. Difference Eq. Appl. **14**(2), 175–190 (2008)
3. Diagana, T., Elaydi, S., Yakubu, A.-A.: Population models in almost periodic environments. J. Difference Equat. Appl. **13**(4), 239–260 (2007)
4. Franke, J.E., Yakubu, A.-A.: Population models with periodic recruitment functions and survival rates. J. Diff. Equat. Appl. **11**(14), 1169–1184 (2005)

Chapter 7
Piecewise-Defined Difference Equations: Open Problem

Candace M. Kent

Abstract We consider difference equations of the form

$$x_{n+1} = f_n(x_n, x_{n-1}, \ldots, x_{n-k}), \quad n = 0, 1, \ldots,$$

where $k \in \{0, 1, \ldots\}$, f_n is piecewise defined and $f_n : D^{k+1} \to D$, $D \subset \mathbf{R}$, whose behavior of solutions is limited to that of being either eventually periodic or unbounded. There exist numerous examples of difference equations that are both piecewise defined and characterized by having every solution either eventually periodic or unbounded. We briefly describe four such cases. However, not all piecewise-defined difference equations have solutions with this behavior, and we point out some of these exceptions. We then present some properties that our sampling of eventually periodic or unbounded piecewise-defined difference equations have in common. We follow up with an open problem, asking for an explanation as to why certain piecewise-defined difference equations have eventually periodic or unbounded solutions, and others do not.

Keywords Piecewise-defined difference equations • Eventually periodic solutions • Unbounded solutions • Max-type equations • Collatz-type equations • Neuronic models • Tent map

7.1 Introduction and Preliminaries

We give a brief synopsis of four well-known cases in the literature of piecewise-defined difference equations characterized by their having every solution eventually

C.M. Kent (✉)
Department of Mathematics and Applied Mathematics, Virginia Commonwealth University,
Grace E. Harris Hall, 4th Floor, 1015 Floyd Avenue, Richmond, Virginia 23284-2014, USA
e-mail: cmkent@vcu.edu

B. Toni et al. (eds.), *Bridging Mathematics, Statistics, Engineering and Technology*,
Springer Proceedings in Mathematics & Statistics 24, DOI 10.1007/978-1-4614-4559-3_7,
© Springer Science+Business Media New York 2012

periodic, every solution unbounded, or every solution either eventually periodic or unbounded. (See, e.g., [1–5, 7–18, 19a, 21–24, 26–31], and the references therein.)

We first define what it means to be periodic, eventually periodic, and unbounded.

Definition 7.1.1. Let $\{x_n\}_{n=-k}^{\infty}$ be a solution of the difference equation

$$x_{n+1} = f_n(x_n, x_{n-1}, \ldots, x_{n-k}), \quad n = 0, 1, \ldots,$$

where $k \in \{0, 1, \ldots\}$ and $f_n : D^{k+1} \to D, D \subset \mathbf{R}$. Then $\{x_n\}_{n=-k}^{\infty}$ is said to be *periodic with period p* if

$$x_{n+p} = x_n \quad \text{for all } n \geq -k.$$

A solution which is said to be periodic with *prime period p* is periodic with period p but not for any value less than p.

Definition 7.1.2. Let $\{x_n\}_{n=-k}^{\infty}$ be a solution of the difference equation

$$x_{n+1} = f_n(x_n, x_{n-1}, \ldots, x_{n-k}), \quad n = 0, 1, \ldots,$$

where $k \in \{0, 1, \ldots\}$ and $f_n : D^{k+1} \to D, D \subset \mathbf{R}$. Then $\{x_n\}_{n=-k}^{\infty}$ is said to be *eventually periodic with period p* (*or truncated periodic*) if there is $N \geq -k$ such that

$$x_{n+p} = x_n \quad \text{for all } n \geq N.$$

Definition 7.1.3. Let $\{x_n\}_{n=-k}^{\infty}$ be a solution of the difference equation

$$x_{n+1} = f_n(x_n, x_{n-1}, \ldots, x_{n-k}), \quad n = 0, 1, \ldots,$$

where $k \in \{0, 1, \ldots\}$ and $f_n : D^{k+1} \to D, D \subset \mathbf{R}$. Then $\{x_n\}_{n=-k}^{\infty}$ is said to be *bounded* if there is $M > 0$ such that

$$|x_n| \leq M \quad \text{for all } n \geq -k.$$

Therefore, $\{x_n\}_{n=-k}^{\infty}$ is *unbounded* if there exists a subsequence $\{x_{n_i}\}_{i=0}^{\infty}$ such that

$$\lim_{i \to \infty} x_{n_i} = +\infty \text{ or } -\infty.$$

The four cases that we consider in this paper are the following.

Case 1. Max-Type Equation

This category includes equations of the form

$$x_{n+1} = \max\left\{\frac{A_n^{(0)}}{x_n}, \frac{A_n^{(1)}}{x_{n-1}}, \ldots, \frac{A_n^{(k)}}{x_{n-k}}\right\}, \quad n = 0, 1, \ldots,$$

where initial conditions, $x_{-k}, x_{-k+1}, \ldots, x_{-1}, x_0$, are positive and the coefficients (also called parameters), $\{A_n^{(i)}\}_{n=0}^{\infty}$, for $i = 0, 1, \ldots, k$, are positive periodic sequences with respective periods $p_i \in \{1, 2, \ldots\}$. When $p_i = 1$ for all $i = 0, 1, \ldots, k$ (i.e., all coefficients are constant), the equation is referred to as *autonomous*; otherwise the equation is referred to as *nonautonomous*.

Case 2. Collatz-Type Equations
   Included under this category are equations of the form

$$x_{n+1} = \begin{cases} \dfrac{\alpha x_n + \beta x_{n-1}}{2}, & \text{if } x_n + x_{n-1} \text{ is even,} \\ \gamma x_n + \delta x_{n-1}, & \text{if } x_n + x_{n-1} \text{ is odd,} \end{cases} \quad n = 0, 1, \ldots,$$

where initial conditions, x_{-1}, x_0, are integers and $\alpha, \beta, \gamma, \delta \in \{-1, 1\}$.

Case 3. Neuronic Models
   The models considered are

(i) The equations

$$x_{n+1} = x_n - g(x_{n-k}), \quad n = 0, 1, \ldots,$$

   and

$$x_{n+1} = x_n + g(x_{n-k}), \quad n = 0, 1, \ldots,$$

   where the delay $k \in \{0, 1, \ldots\}$ and

$$g(u) = \begin{cases} -1, & \text{if } u \le \sigma, \\ 1, & \text{if } u > \sigma, \end{cases}$$

   with threshold $\sigma \in \mathbf{R}$
(ii) The system

$$\begin{cases} x_{n+1} = x_n + ag(x_{n-k}) + bg(y_{n-k}), \\ y_{n+1} = y_n - bg(x_{n-k}) + ag(y_{n-k}), \end{cases} \quad n = 0, 1, \ldots,$$

   where the delay $k \in \{0, 1, \ldots\}$ and

$$g(u) = \begin{cases} -1, & \text{if } u > 0 \\ 1, & \text{if } u \le 0 \end{cases}$$

The following definition, modified from [20], will be useful in the sequel.

Definition 7.1.4. Let σ be some threshold value, and let $\{x_n\}_{n=-k}^{\infty}$ be a solution of the difference equation

$$x_{n+1} = f_n(x_n, x_{n-1}, \ldots, x_{n-k}), \quad n = 0, 1, \ldots,$$

where $k \in \{0,1,\ldots\}$ and $f_n : D^{k+1} \to D$, $D \subset \mathbf{R}$. A *positive semicycle* of $\{x_n\}_{n=-k}^{\infty}$ is a "string" of terms $\{x_l, x_{l+1}, \ldots, x_m\}$, all greater than or equal to σ, with $l \geq -k$ and $m \leq \infty$ and such that

$$\text{either } l = -k \text{ or } l > -k \text{ and } x_{l-1} < \sigma$$

and

$$\text{either } m = \infty \text{ or } m < \infty \text{ and } x_{m+1} < \sigma.$$

A *negative semicycle* of $\{x_n\}_{n=-k}^{\infty}$ is a "string" of terms $\{x_l, x_{l+1}, \ldots, x_m\}$, all less than σ, with $l \geq -k$ and $m \leq \infty$ and such that

$$\text{either } l = -k \text{ or } l > -k \text{ and } x_{l-1} \geq \sigma$$

and

$$\text{either } m = \infty \text{ or } m < \infty \text{ and } x_{m+1} \geq \sigma.$$

Case 4. The Tent Map

The tent map T is the piecewise linear map

$$T(x) = \begin{cases} \alpha x, & \text{if } 0 \leq x \leq \dfrac{1}{\alpha}, \\ \dfrac{\alpha}{\alpha - 1}(1 - x), & \text{if } \dfrac{1}{\alpha} < x \leq 1, \end{cases}$$

where $T : [0,1] \to [0,1]$ and $\alpha \in (1, \infty)$. This map can also be written in the form of a difference equation:

$$x_{n+1} = \begin{cases} \alpha x_n, & \text{if } 0 \leq x_n \leq \dfrac{1}{\alpha}, \\ \dfrac{\alpha}{1 - a}(1 - x_n), & \text{if } \dfrac{1}{\alpha} < x_n \leq 1, \end{cases}$$

where the initial condition $x_0 \in [0,1]$. The following definitions and note, taken from [6, 19], and [25], will be useful in the sequel.

Definition 7.1.5. A complex number α is called an *algebraic number* if it satisfies an equation of the form

$$\alpha^m + a_{m-1}\alpha^{m-1} + \cdots + a_1\alpha + a_0 = 0$$

with rational coefficients $a_0, \ldots, a_{m-1}$. A complex number α is an *algebraic integer* if it satisfies such an equation with integer coefficients. The roots other than α are called the *Galois conjugates of α*.

Definition 7.1.6. A *Pisot number* is a real algebraic integer greater than one whose Galois conjugates all have modulus strictly less than one.

Note 7.1.7. An *almost integer* is a number "very close to an integer" (which is frequently coincidental although not always). For example,

$$e^\pi - \pi = 19.999099791\ldots$$

is an almost integer. In particular, we have that Pisot numbers generate almost integers, and this is not coincidental. They do this as follows: Let α be a Pisot number. Then the values of successively higher powers of α get successively "closer" to whole numbers. What this means is that if one defines, for $x \in \mathbf{R}$,

$$\| x \| = |x - m| \text{ such that } m \text{ is the closest integer to } x,$$

then

$$\lim_{n \to \infty} \| \alpha^n \| = 0.$$

We state the following result which will be pertinent in the sequel.

Proposition 7.1.8. *Let* $\{x_n\}_{n=-k}^\infty$ *be a solution of the difference equation*

$$x_{n+1} = f_n(x_n, x_{n-1}, \ldots, x_{n-k}), \quad n = 0, 1, \ldots,$$

where $k \in \{0, 1, \ldots\}$ *and* $f_n : \mathbf{Z}^{k+1} \to \mathbf{Z}$. *Then either* $\{x_n\}_{n=-k}^\infty$ *is unbounded or* $\{x_n\}_{n=-k}^\infty$ *is eventually periodic.*

We end this section with what might be considered one of the simplest piecewise-linear difference equations whose every solution is eventually periodic:

$$x_{n+1} = |x_n - 1|, \quad n = 0, 1, \ldots,$$

where the initial condition $x_0 \in \mathbf{R}$.

7.2   Max-Type Equations

In a key paper by Amleh et al. [2], it was shown that every solution of the difference equation

$$x_{n+1} = \max\left\{ \frac{1}{x_n}, \frac{A}{x_{n-1}} \right\}, \quad n = 0, 1, \ldots, \tag{7.1}$$

where the constant coefficient A and initial conditions x_{-1}, x_0 are nonzero real numbers, is eventually periodic. One of the results obtained was that when A and x_{-1}, x_0 are positive, every positive solution $\{x_n\}_{n=-1}^\infty$ of Eq. (7.1) is eventually periodic with period

1. Two if $A \in (0, 1)$
2. Three if $A = 1$
3. Four if $A \in (1, \infty)$

The proof given was geometric and showed how in a *finite* number of iterations any point (x_{n-1}, x_n) in the x_{n-1}, x_n phase plane "jumps" onto a particular line segment along which all solutions are periodic.

On the other hand, with the equation

$$x_{n+1} = \max\left\{\frac{a}{x_n}, \frac{A}{x_{n-1}}\right\}, \quad n = 0, 1, \ldots,$$

where $a \neq A$, a, A are negative, and x_{-1}, x_0 are nonzero real numbers, it was found that every solution is unbounded.

Briden et al. [5] made the coefficient A in Eq. (7.1) variable and investigated the resulting nonautonomous difference equation

$$x_{n+1} = \max\left\{\frac{1}{x_n}, \frac{A_n}{x_{n-1}}\right\}, \quad n = 0, 1, \ldots, \tag{7.2}$$

where $\{A_n\}_{n=0}^{\infty}$ is a periodic sequence of positive real numbers with period two such that

$$A_n = \begin{cases} A_0, & \text{if } n \text{ is even}, \\ A_1, & \text{if } n \text{ is odd}, \end{cases}$$

and where the initial conditions x_{-1}, x_0 are positive. It was shown that every positive solution $\{x_n\}_{n=-1}^{\infty}$ of Eq. (7.2) is eventually periodic with the following periods:

1. Two if $A_0 A_1 \in (0, 1)$
2. Six if $A_0 A_1 = 1$
3. Four if $A_0 A_1 \in (1, \infty)$

Briden et al. [4] and Grove et al. [15] then studied this same equation

$$x_{n+1} = \max\left\{\frac{1}{x_n}, \frac{A_n}{x_{n-1}}\right\}, \quad n = 0, 1, \ldots, \tag{7.3}$$

only now with the positive sequence $\{A_n\}_{n=0}^{\infty}$ periodic with period three such that, for $m \geq 0$,

$$A_n = \begin{cases} A_0, & \text{if } n = 3m, \\ A_1, & \text{if } n = 3m + 1, \\ A_2, & \text{if } n = 3m + 2. \end{cases}$$

Initial conditions, x_{-1}, x_0, were kept positive so that again all solutions, $\{x_n\}_{n=-1}^{\infty}$, were positive. The following was proved:

1. If $A_n \in (0, 1)$ for all $n \geq 0$, then every positive solution of Eq. (7.3) is eventually periodic with period two.
2. If $A_n \in (1, \infty)$ for all $n \geq 0$, then every positive solution of Eq. (7.3) is eventually periodic with period twelve.

3. If $A_{i+1} < 1 < A_i$ for some $i \in \{0,1,2\}$, then every positive solution of Eq. (7.3) is unbounded.
4. In all other cases, every positive solution of Eq. (7.3) is eventually periodic with period three.

The proofs in the above two cases were purely computational and inductive. Noteworthy is the fact that the branching of possibilities in each proof is curtailed since a *finite* set of conditions evolves with each iteration that then completely determines which argument of the maximum function is chosen at each iterative step.

In a paper by Kent and Radin [18], the following equation was examined:

$$x_{n+1} = \max \left\{ \frac{A_n}{x_n}, \frac{B_n}{x_{n-1}} \right\}, \quad n = 0,1,\ldots, \tag{7.4}$$

where $\{A_n\}_{n=0}^{\infty}$ is a periodic sequence of positive real numbers with prime period p, and $\{B_n\}_{n=0}^{\infty}$ is a periodic sequence of positive real numbers with prime period q. They discovered that every positive solution, $\{x_n\}_{n=-1}^{\infty}$, of Eq. (7.4) is unbounded if p or q is a multiple of three. A number-theoretic approach was taken in the proof. All previous results cited were extended in yet another landmark paper by Bidwell and Franke [3]. It was shown that every *bounded* positive solution $\{x_n\}_{n=-k}^{\infty}$ of the equation

$$x_{n+1} = \max \left\{ \frac{A_n^{(0)}}{x_n}, \frac{A_n^{(1)}}{x_{n-1}}, \ldots, \frac{A_n^{(k)}}{x_{n-k}} \right\}, \quad n = 0,1,\ldots, \tag{7.5}$$

where $k \in \{1,2,\ldots\}$ and $\left\{A_n^{(i)}\right\}_{n=0}^{\infty}$, for $i = 0,1,\ldots,k$, is a periodic sequence of nonnegative real numbers with period $p_i \in \{1,2,\ldots\}$, is eventually periodic. It is interesting to note that the proof here shares some common features with the proof used in [2]. Both proofs use a log transformation, which converts the equation into a piecewise-linear equation, and both involve points (x_n, x_{n-1}) asymptotically approaching but then suddenly "jumping" onto a set of points representing solutions that are periodic. With Eq. (7.5), the set is an ω-limit cycle.

In contrast to the above examples, there are piecewise-defined difference equations with the maximum function whose solutions are not necessarily eventually periodic but asymptotically convergent to some value.

For example, in a paper by Stevic [29], it was proved that every positive solution of the difference equation

$$x_{n+1} = \max \left\{ \frac{A_0}{x_n^{\alpha_0}}, \frac{A_1}{x_{n-1}^{\alpha_1}}, \ldots, \frac{A_k}{x_{n-k}^{\alpha_k}} \right\}, \quad n = 0,1,\ldots, \tag{7.6}$$

where $k \in \{1,2,\ldots\}$ and $A_i \in (0,\infty)$ and $\alpha_i \in (0,1)$, for $i = 0,1,\ldots,k$, converges to

$$\bar{x} = \max \left\{ A_0^{\frac{1}{\alpha_0+1}}, \ldots, A_k^{\frac{1}{\alpha_k+1}} \right\}.$$

In a paper by Yang et al. [30], mixed results were obtained with an example of Eq. (7.6),

$$x_{n+1} = \max\left\{\frac{1}{x_n^\alpha}, \frac{A}{x_{n-1}}\right\}, \quad n = 0, 1, \ldots,$$

where $A \in (0, \infty)$ and $\alpha \in (0, 1)$. Their results were as follows:

1. If $A \in (1, \infty)$, then every positive solution is eventually periodic with period four. In particular, there exists $N \geq 0$ such that $x_n = \frac{A}{x_{n-1}}$ for all $n \geq N$.
2. If $A \in (0, 1]$, then every positive solution converges to $\bar{x} = 1$.

Recently Sauer in [26] and [27] generalized the idea of max-type equations to that of *rank-type equations*. This new class of equations includes, for example, equations of the form

$$x_{n+1} = \max\{f_0(x_n), f_1(x_{n-1}), \ldots, f_k(x_{n-k})\}, \quad n = 0, 1, \ldots,$$

$$x_{n+1} = \min\{f_0(x_n), f_1(x_{n-1}), \ldots, f_k(x_{n-k})\}, \quad n = 0, 1, \ldots,$$

$$x_{n+1} = \text{median}\{f_0(x_n), f_1(x_{n-1}), \ldots, f_k(x_{n-k})\}, \quad n = 0, 1, \ldots,$$

where $k \in \{1, 2, \ldots\}$ and $f_i : \mathbf{R} \to \mathbf{R}$, for $i \in \{0, 1, \ldots\}$, is continuous. A function is called *contractive* if there exists $\alpha \in [0, 1)$ and a real number r such that $|f(x) - r| \leq \alpha|x - r|$ for all x. With, for example, the max equation just above, Sauer showed that if f_i, for $i = 0, 1, \ldots, k$, is contractive (i.e., for $i \in \{0, 1, \ldots\}$, there exist $\alpha_i \in [0, 1)$ and $r_i \in \mathbf{R}$ such that $|f_i(x) - r_i| \leq \alpha_i|x - r_i|$ for all x), then every solution $\{x_n\}_{n=0}^\infty$ converges to $\max\{r_0, r_1, \ldots, r_k\}$. However we make mention of the fact that if $\alpha < 1$ is relaxed, then, in many instances, the solution is eventually periodic rather than asymptotically convergent. A typical instance is with our familiar equation

$$x_{n+1} = \max\left\{\frac{A_n^{(0)}}{x_n}, \frac{A_n^{(1)}}{x_{n-1}}, \ldots, \frac{A_n^{(k)}}{x_{n-k}}\right\}, \quad n = 0, 1, \ldots.$$

Given all of the above results, we conclude this section with the following observation.

In the case of max-type equations whose solutions are either eventually periodic or unbounded, as against, say, asymptotically convergent, the terms of the solutions are "made up" of elements that are drawn from a *finite* set of elements in much the same way as the nonnegative integers are formed from the combination of only ten digits. Specifically, consider the nonautonomous equation

$$x_{n+1} = \max\left\{\frac{A_n}{x_n}, \frac{B_n}{x_{n-1}}\right\}, \quad n = 0, 1, \ldots,$$

where initial conditions x_{-1}, x_0 are positive and where the variable coefficients $\{A_n\}_{n=0}^{\infty}$, $\{B_n\}_{n=0}^{\infty}$ are both periodic sequences of positive real numbers with prime period two such that

$$A_n = \begin{cases} A_0, & \text{if } n \text{ is even,} \\ A_1, & \text{if } n \text{ is odd,} \end{cases}$$

$$B_n = \begin{cases} B_0, & \text{if } n \text{ is even,} \\ B_1, & \text{if } n \text{ is odd.} \end{cases}$$

Then the "finite set of elements" is $\{x_{-1}, x_0, A_0, A_1, B_0, B_1\}$ and every term x_n of a solution $\{x_n\}_{n=-1}^{\infty}$ is equal to the countable product of these elements:

$$x_n = x_{-1}^{n_1} \cdot x_0^{n_2} \cdot A_0^{n_3} \cdot A_1^{n_4} \cdot B_0^{n_5} \cdot B_1^{n_6},$$

where $n_i \in \mathbf{Z}$ for $i \in \{1, 2, 3, 4, 5, 6\}$

7.3 Collatz-Type Equations

In 1950, Lothar Collatz introduced what is currently called the $3x + 1$ *Problem*, associated with the map

$$C(x) = \begin{cases} \dfrac{x}{2}, & \text{if } x \equiv 0 \pmod 2, \\ 3x + 1, & \text{if } x \equiv 1 \pmod 2, \end{cases}$$

with $x \in \mathbf{N}$, at the International Congress of Mathematicians in Cambridge, Massachusetts. This map is now referred to as the *Collatz map*. The problem was to prove the conjecture that for any $x \in \mathbf{N}$, there exists $N \geq 0$ such that $C^N(x) = 1$, after which iterations cycle through the values 4, 2, 1. (See Lagarias [22].) This problem has averted solution up to the present.

The Collatz map can be written more concisely as

$$T(x) = \begin{cases} \dfrac{x}{2}, & \text{if } x \equiv 0 \pmod 2, \\ \dfrac{3x + 1}{2}, & \text{if } x \equiv 1 \pmod 2, \end{cases}$$

where $x \in \mathbf{N}$ and T is called the $3x + 1$ *map*. We can, in turn, translate this map into the difference equation (which we will refer to as the "$3x + 1$ difference equation")

$$x_{n+1} = \begin{cases} \dfrac{x_n}{2}, & \text{if } x_n \text{ is even,} \\ \dfrac{3x_n + 1}{2}, & \text{if } x_n \text{ is odd,} \end{cases} \quad n = 0, 1, \dots, \tag{7.7}$$

where $x_0 \in \mathbf{N}$; and we can restate the $3x + 1$ Conjecture as follows:

Conjecture 7.3.1. Every solution of Eq. (7.7) is eventually periodic with period two, ending in the two-cycle $(1,2)$.

Since 1950, there have been partial proofs of the conjecture or of related conjectures, and variants of the $3x + 1$ equation together with the creation of *Collatz-type equations*, more amenable to solution, have sprouted up.

We consider the Collatz-type equations.

One of the earlier investigations of these Collatz-type difference equations was conducted by Clark and Lewis [9]. They studied the equation

$$x_{n+1} = \begin{cases} \dfrac{x_n + x_{n-1}}{2}, & \text{if } x_n + x_{n-1} \text{ is even,} \\ x_n - x_{n-1}, & \text{if } x_n + x_{n-1} \text{ is odd,} \end{cases} \qquad n = 0, 1, \ldots, \qquad (7.8)$$

where $x_{-1}, x_0 \in \mathbf{Z}$. Clark and Lewis' goal was to show that indeed every solution is eventually periodic. In order to do so, they needed to find a bound on each solution which would be a function of the initial conditions x_{-1}, x_0 (see Proposition 7.1.8). They found such a bound:

$$|x_n| \le \max\{|x_{-1}|, |x_0|\} \quad \text{for all } n \ge -1.$$

They then not only were able to conclude that every solution is eventually periodic, but, using this bound, were able to show that a solution $\{x_n\}_{n=-1}^{\infty}$ is either eventually the constant solution $1, 1, \ldots$, the constant solution $-1, -1 \ldots$, or the period-six solution ending in the six-cycle $(-2, 1, 3, 2, -1, -3)$.

Clark and Lewis' equation was then extended to the following family of sixteen Collatz-type equations by Al-Amleh et al. [1]:

$$x_{n+1} = \begin{cases} \dfrac{\alpha x_n + \beta x_{n-1}}{2}, & \text{if } x_n + x_{n-1} \text{ is even,} \\ \gamma x_n + \delta x_{n-1}, & \text{if } x_n + x_{n-1} \text{ is odd,} \end{cases} \qquad n = 0, 1, \ldots, \qquad (7.9)$$

where $x_{-1}, x_0 \in \mathbf{Z}$ and $\alpha, \beta, \gamma, \delta \in \{-1, 1\}$. The solutions of all but four of the sixteen equations were fully characterized and were found to be either eventually periodic or unbounded. For example, when $\alpha = \beta = \gamma = \delta = 1$, it was proved that every solution is eventually constant (periodic with period one) or unbounded. When it came to the four unsolved equations, it was conjectured that with each, every solution is eventually periodic. However, no bounds on solutions were discovered (as has been the case with the $3x + 1$ Problem), thereby making it seemingly impossible to prove the conjectures.

Feuer and Ladas in a series of two papers [13] and [11] proposed several open problems and conjectures on Collatz-type equations that are generalizations of Eq. (7.9):

1. They looked at six cases of the family of equations

$$x_{n+1} = \begin{cases} \dfrac{\alpha x_n + \beta x_{n-1}}{a}, & \text{if } a \,|\, x_n + x_{n-1}, \\ \gamma x_n + \delta x_{n-1}, & \text{otherwise}, \end{cases} \quad n = 0, 1, \ldots, \quad (7.10)$$

where $x_{-1}, x_0 \in \mathbf{Z}$, a is an integer greater than or equal to 2, and $\alpha, \beta, \gamma, \delta \in \{-1, 1\}$.

2. Then they studied eight cases of the family of equations

$$x_{n+1} = \begin{cases} \dfrac{\alpha x_n + \beta x_{n-1}}{a_n}, & \text{if } a_n \,|\, x_n + x_{n-1}, \\ \gamma x_n + \delta x_{n-1}, & \text{otherwise}, \end{cases} \quad n = 0, 1, \ldots, \quad (7.11)$$

where $x_{-1}, x_0 \in \mathbf{Z}$, $\{a_n\}_{n=0}^{\infty}$ is a periodic sequence of integers greater than or equal to 2, and $\alpha, \beta, \gamma, \delta \in \{-1, 1\}$.

It was conjectured that in some of the cases, every solution is eventually periodic, and in others, unbounded.

The proofs used in the above examples are inductive in nature and case-specific. In a recent paper by Liddell [24], an alternative geometric approach was suggested that could serve as a more general method for analyzing piecewise-linear difference equations (especially those, for example, represented by Eq. (7.10)) in the phase space $\mathbf{Z}^2$.

While it is obvious why every solution of a Collatz-type equation, for which bounds on solutions have been discovered, is eventually periodic, what about the actual $3x + 1$ difference equation and the four unsolved cases of Eq. (7.9), where no bounds have been obtained? Computer observation tells us that in these latter cases, every solution is eventually periodic, which means there have to be bounds on these integer solutions. But how do we find such bounds? Do we imbed these problems in larger problems, for which the mathematics may not yet be known, and look for the bounds there? As Paul Erdös stated, "Mathematics is not yet ready for such problems." (See Lagarias [22].)

7.4 Neuronic Models

There is a paucity of applications of difference equations to the field of neuroscience, especially where analytic proofs are involved rather than simply computer observations and simulations. A few of elegant examples exist in which it is proved that the solutions of the difference equation models are eventually periodic or unbounded. The difference equations are all piecewise constant with thresholds.

In a paper by Chen [7], the following difference equation model was presented, which was a discretized version of a differential equation with applications in neural network theory:

$$x_{n+1} = x_n - g(x_{n-k}), \quad n = 0, 1, \ldots, \tag{7.12}$$

where $k \in \{0, 1, \ldots\}$ and

$$g(u) = \begin{cases} -1, & \text{if } u \leq \sigma, \\ 1, & \text{if } u > \sigma, \end{cases}$$

with *threshold value* $\sigma \in \mathbf{R}$. Chen proved that every solution of Eq. (7.12) is eventually periodic with prime period $2(2l + 1)$ for some $l \geq 0$ such that $\frac{k-l}{2l+1}$ is a nonnegative integer.

In another paper by Chen [8], it was shown that every solution of a modified version of Eq. (7.12),

$$x_{n+1} = x_n + g(x_{n-k}), \quad n = 0, 1, \ldots,$$

is either eventually periodic or unbounded (unbounded if the solution has a semicycle with length greater than k–See the Introduction for the definition of a "semicycle").

Yuan and Huang [31] extended Chen's work to the following system:

$$x_{n+1} = x_n + ag(x_{n-k}) + bg(x_{n-k}),$$
$$\qquad\qquad\qquad\qquad\qquad n = 0, 1, \ldots$$
$$y_{n+1} = y_n - bg(x_{n-k}) + ag(y_{n-k}),$$

where a, b are constants, $k \in \{0, 1, \ldots\}$, and

$$g(u) = \begin{cases} -1, & \text{if } u > 0, \\ 1, & \text{if } u \leq 0, \end{cases}$$

Every solution, $\{(x_n, y_n)\}_{n=-k}^{\infty}$, was again shown to be eventually periodic, but not unbounded.

The proofs in the first two papers utilize *semicycle analysis* in which semicycles of terms of solutions are studied in relation to the threshold value σ. The fact that the difference equations involve thresholds suggests that some sort of feedback mechanisms in the mathematical sense are at work; namely, negative feedback leading to convergence of solutions to eventually periodic solutions and positive feedback giving rise to unbounded solutions. In addition, again, as with max-type equations the terms of the solutions are "made up" of elements that are drawn from the *finite* set of elements $\{x_0, -1, 1\}$, where each term is of the form $x_0 + n$, $n \in \mathbf{Z}$ (see Proposition A). The proofs in the third paper are both geometric in nature and in the spirit of Proposition A.

7.5 The Tent Map

We introduce the *tent map*, a piecewise linear map, with the following example (see [10]):

$$T(x) = \begin{cases} 2x, & \text{if } 0 \leq x \leq \dfrac{1}{2}, \\[2mm] 2(1-x), & \text{if } \dfrac{1}{2} < x \leq 1, \end{cases} \tag{7.13}$$

where $T : [0,1] \to [0,1]$. If the map T is written as a difference equation we have

$$x_{n+1} = \begin{cases} 2x_n, & \text{if } 0 \leq x_n \leq \dfrac{1}{2}, \\[2mm] 2(1-x_n), & \text{if } \dfrac{1}{2} < x_n \leq 1. \end{cases} \tag{7.14}$$

For the most part, this example exhibits chaotic behavior. However, if the initial condition $x_0 = \frac{k}{2^m} \in (0,1)$ with $k, m \in \mathbf{N}$, then the resulting solution $\{x_n\}_{n=0}^{\infty}$ of Eq. (7.14) is eventually constant (i.e., periodic with period one).

The map in Eq. (7.13) is just one member of the family of *asymmetric tent maps* (see [23]):

$$T_\alpha(x) = \begin{cases} \alpha x, & \text{if } 0 \leq x \leq \dfrac{1}{\alpha}, \\[2mm] \dfrac{\alpha}{\alpha - 1}(1-x), & \text{if } \dfrac{1}{\alpha} < x \leq 1, \end{cases}$$

where $T_\alpha : [0,1] \to [0,1]$ for $\alpha \in (1,\infty)$. The corresponding difference equation of this map is

$$x_{n+1} = \begin{cases} \alpha x_n, & \text{if } 0 \leq x_n \leq \dfrac{1}{\alpha}, \\[2mm] \dfrac{\alpha}{\alpha - 1}(1-x_n), & \text{if } \dfrac{1}{\alpha} < x_n \leq 1. \end{cases} \tag{7.15}$$

It is known that for each α, there exists a subset of initial conditions in the set $\mathbf{Q}(\alpha) \cap [0,1]$, where $\mathbf{Q}(\alpha)$ is a finite algebraic extension of $\mathbf{Q}$, such that the resulting solutions are eventually periodic. However, if α belongs to a "special subgroup" of the Pisot numbers (with this subgroup, $\frac{\alpha}{\alpha-1}$ is also a Pisot number), then *all* initial conditions in the set $\mathbf{Q}(\alpha) \cap [0,1]$ give rise to eventually periodic solutions. (See [6, 19, 23], and also the definition of "Pisot number" in the Introduction.) In addition, we reiterate that high powers of Pisot numbers are almost integers (see [25] and also the definition of "almost integers in the Introduction).

Therefore, in Eq. (7.15), when α belongs to this special subgroup of Pisot numbers, referred to above, then there is a subset of (almost) integers that go into "making up" the terms x_n of solutions $\{x_n\}_{n=0}^{\infty}$ of Eq. (7.15). Couple this with the fact that every solution is bounded, that is, $x_n \in [0,1]$, for all $n = 0, 1, \ldots$, and we have "almost" Proposition A.

7.6 Open Problem, Heuristic Argument, and Query

We conclude with our open problem, a brief heuristic argument, and a suggestion on using an adjuvant approach to the study of piecewise-defined difference equations.

7.6.1 Open Problem

Open Problem 1. Develop a complete set of properties which allows one to distinguish between piecewise-defined difference equations whose solutions are either eventually periodic or unbounded and piecewise-defined and other difference equations that do not exhibit this type of behavior.

It should be noted that E.A. Grove and G. Ladas in their book, *Periodicities in Nonlinear Difference Equations* [16], inspired the formulation of this open problem.

7.6.2 Heuristic Argument

Observe that each of the piecewise-defined difference equations considered above involves some type of threshold. For example, with the max-type equation

$$x_{n+1} = \max\left\{\frac{1}{x_n}, \frac{A}{x_{n-1}}\right\}, \quad n = 0, 1, \ldots,$$

we have A as our threshold value if we rewrite the equation as

$$x_{n+1} = \begin{cases} \dfrac{1}{x_n}, & \text{if } \dfrac{x_{n-1}}{x_n} \geq A, \\[3mm] \dfrac{A}{x_n}, & \text{if } \dfrac{x_{n-1}}{x_n} \leq A, \end{cases} \quad n = 0, 1, \ldots.$$

We can rewrite the $3x + 1$ difference equation as

$$x_{n+1} = \begin{cases} \dfrac{x_n}{2}, & \text{if } x_n \equiv 0 \pmod 2, \\[3mm] \dfrac{3x_n + 1}{2}, & \text{if } x_n \equiv 1 \pmod 2. \end{cases} \quad n = 0, 1, \ldots.$$

Here, the threshold value can be viewed as 0 (mod 2), where either x_n is either equal to 0 (mod 2) or is "greater than" 0 (mod 2).

The neuronic models

$$x_{n+1} = x_n \pm g(x_{n_k}), \quad n = 0, 1, \ldots,$$

where $k \in \{0, 1, \ldots\}$ and

$$g(u) = \begin{cases} -1, & \text{if } u \leq \sigma, \\ 1, & \text{if } u > \sigma, \end{cases}$$

have the already built-in threshold value $\sigma \in \mathbf{R}$.

Perhaps there is some type of negative feedback going on in which there is convergence, albeit, not asymptotic, with solutions eventually becoming periodic; or some type of positive feedback going on in which there is divergence or unboundedness of solutions. If negative feedback and positive feedback do in actuality exist, they need to be characterized rigorously. The fact that convergence to periodic solutions is not asymptotic suggests that Proposition A is somehow playing a role in solutions eventually becoming periodic.

7.6.3 Query

Difference equations today have many applications to areas outside of mathematics, including biology and medicine. What about an application in the opposite direction? There are cases where biology has motivated ideas in mathematics. In particular, there is a paper, relevant to our paper, by Sayama [28], in which morphogenesis (the study of the evolution and development of shapes and patterns of organisms) is used to aid in the elucidation of the $3x + 1$ problem. Perhaps doing something similar in the case of Open Problem 7.10, would provide at least part of its solution.

References

1. Al-Amleh, A., Grove, E.A., Kent, C.M., Ladas, G.: On some difference equations with eventually periodic solutions. J. Math. Anal. Appl. **233**, 196–215 (1998)
2. Amleh A.M., Hoag, J., Ladas, G.: A difference equation with eventually periodic solutions. Comput. Math. Appl. **36**, 401–404 (1998)
3. Bidwell, J., Franke, John, E.: Bounded implies eventually periodic for the positive case of reciprocal-max difference equation with periodic parameters. J. Differ. Equat. Appl. **14**(3), 321–326 (2008)
4. Briden, W.J., Grove, E.A., Kent, C.M., Ladas, G.: Eventually periodic solutions of the difference equation $x_{n+1} = \max\{\frac{1}{x_n}, \frac{A_n}{x_{n-1}}\}$. Commun. Appl. Nonliear Anal. **6**(4), 31–43 (1999)
5. Briden, W.J., Grove, E.A., Ladas, G., McGrath, L.C.: On the nonautonomous equation $x_{n+1} = \max\{\frac{A_n}{x_n}, \frac{B_n}{x_{n-1}}\}$. In: Proceedings of the Third International Conference on Difference Equations and Applications, 1–5 Sept 1997, pp 49–73. Gordon and Breach Science Publishers, Taipei, Taiwan (1999)

6. Burr, S.A. (ed.): The Unreasonable Effectiveness of Number Theory. In: Proceedings of Symposia in Applied Mathematics, Vol. 46. American Mathematical Society (1992)
7. Chen, Y.: All solutions of a class of difference equations truncated periodic. Appl. Math. Lett. **15**, 975–979 (2002)
8. Chen, Y.: Limiting behavior of a class of delay difference equations. Dynamics of Continuous, Discrete and Impulsive Systems Series A: Math. Anal. **10**, 75–80 (2003)
9. Clark, D., Lewis, J.: A Collatz-type difference equation. Congressus Numeratium. **11**, 129–135 (1995)
10. Elaydi, S.N.: Discrete Chaos. Chapman & Hall/CRC Press (2000)
11. Feuer, J., Ladas, G.: Some equations with eventually periodic solutions. J. Differ. Equat. Appl. **10**(4), 447–451 (2004)
12. Feuer, J.: On the eventual periodicity of $x_{n+1} = \max\{\frac{1}{x_n}, \frac{A_n}{x_{n-1}}\}$. J. Differ. Equat. Appl. **12**, 467–486 (2006)
13. Feuer, J.: Some equations with periodic parameter and eventually periodic solutions. J. Differ. Equat. Appl. **13**(11), 1005–1010 (2007)
14. Gelisken, A., Cinar, C., Kurbanli, Abdullah S.: On the asymptotic behavior and periodic nature of a difference equation with maximum. Comput. Math. Appl. **59**, 898–902 (2010)
15. Grove, E.A., Kent, C., Ladas, G., Radin, M.: On $x_{n+1} = \max\{\frac{1}{x_n}, \frac{A_n}{x_{n-1}}\}$ with a period-three parameter. Fields Inst. Commun. **29**, 161–180 (2001)
16. Grove, E.A., Ladas, G.: Periodicities in Nonlinear Difference Equations. Advances in Discrete Mathematics and Applications, Vol. 4. Chapman & Hall/CRC Press (2005).
17. Kent, C.M., Kustesky, M., Nguyen, A.Q., Nguyen, B.V.: Eventually periodic solutions of $x_{n+1} = \max\{\frac{A_n}{x_n}, \frac{B_n}{x_{n-1}}\}$ when the parameters are two-cycles. Dynamics of Continuous, Discrete and Impulsive Systems, Series A: Math. Anal. **10**, 33–49 (2003)
18. Kent, C.M., Radin, M.A.: On the boundedness nature of positive solutions of the difference equation $x_{n+1} = \max\{\frac{A_n}{x_n}, \frac{B_n}{x_{n-1}}\}$ with periodic parameters. In: Proceedings of the Third International DCDIS Conference on Engineering Applications and Computational Algorithms, 15 May 2003, Guelph, Ontario, Canada. Special Volume of the Dynamics of Continuous, Discrete and Impulsive Systems, Series B: Applications and Algorithms, pp. 11–15, Watam Press (2003)
19. Koch, H.: Number Theory: Algebraic Numbers and Functions. Graduate Studies in Mathematics, Vol. 24. American Mathematical Society (2000)
19a. Kocic, V.L.: Dynamics of a piecewise linear map. In: Proceedings of the International Conference on Differential and Difference Equations and Their Applications, Melbourne, Florida, 1–5 Aug 2005, pp 565–567. Hindawi Publishing Corporation, New York (2006)
20. Kocic, V.L., Ladas, G.: Global Behavior of Nonlinear Difference Equations of Higher Order with Applications. Mathematics and its Applications, Vol. 256. Kluwer Academic Publishers (1993)
21. Ladas, G.: On the recursive sequence $x_{n+1} = \max\{\frac{A_0}{x_n}, \frac{A_1}{x_{n-1}}, \ldots, \frac{A_k}{x_{n-k}}\}$. J. Differ. Equat. Appl. **2**, 339–341 (1996)
22. Lagarias, Jeffrey C. (ed.): The Ultimate Challenge: The $3x + 1$ Problem. American Mathematical Society (2010)
23. Lagarias, J.C., Porta, H.A., Stolarsky, K.B.: Asymmetric tent map expansions. I. Eventually periodic points. J. Lond. Math. Soc. **47**(2), 542–556 (1993)
24. Liddell, G.F.: Piecewise linear difference equations and convexity. J. Differ. Equat. Appl. **18**(1), 139–148 (2012)
25. Panju, M.: A systematic construction of almost integers. Waterloo Math. Rev. I(2), 35–43 (2011)
26. Sauer, T.: Global Convergence of max-type equations. J. Differ. Equat. Appl. **17**(1-2), 1–8 (2011)
27. Sauer, T.: Convergence of rank-type equations. Appl. Math. Comput. **217**, 4540–4547 (2011)
28. Sayama, H.: An Artificial life view of the Collatz problem. Artif. Life. **17**,137–140 (2011)

29. Stevic, S.: Global stability of a difference equation with maximum. Appl. Math. Comput. **210**, 525–529 (2009)
30. Yang, X., Liao, X., Li, C.: On the difference equation with maximum, Appl. Math. Comput. **181**, 1–5 (2006)
31. Yuan, Z., Huang, L.: All solutions of discrete-time systems are eventually periodic. Appl. Math. Comput. **158**, 537–546 (2004)

Chapter 8
Mathematics Behind Microstructures: A Lead to Generalizations of Convexity

Daniel Vasiliu

Abstract We consider a mathematical model aimed at explaining pattern formation in microstructures. Usually such models are also useful for understanding problems of solid–solid phase transitions in material science. Our goal is to analyze the limiting behavior of certain non-linear energy type functionals, with restrictions, from a variational point of view. In order to better understand this problem we develop some generalizations for the notions of rank-one convexity and quasi-convexity and demonstrate their relevance in the context of energy minimizing sequences.

Keywords Microstructures • Rank-one convexity • Quasiconvexity • Restricted lower semicontinuity

Mathematics Subject Classification (1991): 49J45

8.1 Introduction

The idea that microstructures are abundant in nature represents a universally accepted truth. In one of the most notable monographs concerning the mathematical models of nonlinear elasticity and microstructures [29], Müller elaborated on the idea that *the observed microstructure corresponds to minimizers or almost minimizers of the elastic energy.* In the case of an elastic crystal, given a reference configuration $\Omega \subset \mathbb{R}^3$, a deformation $u : \Omega \to \mathbb{R}^3$ requires an elastic energy:

D. Vasiliu (✉)
Department of Mathematics, Christopher Newport University, Newport News,
VA 23606, USA
e-mail: daniel.vasiliu@cnu.edu

B. Toni et al. (eds.), *Bridging Mathematics, Statistics, Engineering and Technology,*
Springer Proceedings in Mathematics & Statistics 24, DOI 10.1007/978-1-4614-4559-3_8,
© Springer Science+Business Media New York 2012

$$I(u) = \int_{\Omega} f(Du)\mathrm{d}x \qquad (8.1)$$

where $f : \mathbb{M}^{3\times3} \to \mathbb{R}$ represents the stored-energy density function that describes the properties of the material. Under the Cauchy–Born rule $f(A)$ is given by the (free) energy per unit volume that is required for an affine deformation $x \mapsto Ax$ of the crystal lattice.

Thus, a problem of significant importance in the Calculus of Variations is to find among all functions $u \in W^{1,p}(\Omega, \mathbb{R}^m)$, with certain prescribed constraints, those which minimize a given functional

$$I(u) = \int_{\Omega} f(x, u(x), Du(x))\mathrm{d}x \qquad (8.2)$$

where $f : \Omega \times \mathbb{R}^m \times \mathbb{M}^{m\times n} \to \mathbb{R}$, $\Omega \subset \mathbb{R}^n$ a bounded domain and Du denotes the gradient of u in the sense of distributions. A direct method of proving existence of minimizers is to find minimizing sequences converging in some topology and check that the functional I is lower semicontinuous in that topology; then in this case the limit would be a minimizer. Therefore it is of special interest in finding necessary and sufficient conditions for the function f such that I defined (8.2) is weakly lower semicontinuous on certain Sobolev space. One "right" candidate for such a condition is the concept of *quasiconvexity* first introduced by Morrey in the early 1950s [26]. According to Morrey a function $f : \mathbb{M}^{m\times n} \to \mathbb{R}$ is *quasiconvex* if

$$\int_{\Omega} f(A + Du(x))\mathrm{d}x \geq |\Omega| f(A)$$

for all $A \in \mathbb{M}^{m\times n}$ and all $u \in C_0^{\infty}(\Omega, \mathbb{R}^m)$.

Acerbi and Fusco [2] proved that under some proper growth condition the weak lower semicontinuity of the functional I given by (8.2) is equivalent to the quasiconvexity condition of f with respect to variable ξ.

The quasiconvexity condition is generally difficult to verify. As a major contribution in understanding this condition we distinguish the work of Ball [4]. He developed the concepts of *rank-one convexity* and *polyconvexity* along with the quasiconvexity emphasizing many interesting facts in the attempt to consecrate a useful sufficient condition for the weak lower semicontinuity. It turns out that rank-one convexity (see definition below), although easier to check, is the weakest among all three conditions. In general rank-one convexity does not imply quasiconvexity (Šverák [38]) but vice versa is always true. However there are particular cases when rank-one convexity is equivalent to quasiconvexity, for example, when f is a quadratic form.

An efficient way to study weakly convergent sequences and the weak lower semicontinuity property for the functional (8.2) is to use the concept of Young measures developed by Tartar [42] following the original idea of Young [47]. Kinderlehrer and Pedregal [20] showed that the homogeneous gradient Young measures are exactly those probability measures that satisfy Jensen's inequality for all quasiconvex functions f i.e.,

$$\int_{\mathbb{M}^{m\times n}} f(\lambda)\mathrm{d}v_x(\lambda) \geq f\left(\int_{\mathbb{M}^{m\times n}} \lambda\mathrm{d}v_x(\lambda)\right)$$

Using the techniques of Young measures, Fonseca and Müller [18] studied the so-called $\mathcal{A}$-quasiconvexity problem and Müller [30] also studied a similar problem without the constant rank condition.

In this paper we study the weak lower semicontinuity of functionals I given by (8.2) along sequences u_k satisfying a projection-type constraint (i.e., dist$(Du_k, \mathcal{L})$ $\rightarrow 0$) for a given linear subspace $\mathcal{L}$ of $\mathbb{M}^{m\times n}$. We show that this problem leads to meaningful generalizations of the rank-one convexity and quasiconvexity concepts. We say that a function $f : \mathcal{L} \rightarrow \mathbb{R}$ is $\mathcal{L}$- *rank one convex* if for any $\lambda \in [0,1]$ and A, $B \in \mathcal{L}$ such that rank$(A - B) \leq 1$ we have

$$f(\lambda A + (1-\lambda)B) \leq \lambda f(A) + (1-\lambda)f(B).$$

Also we say that f is $\mathcal{L}$- *quasiconvex* if

$$f(A) \leq \frac{1}{|Q|}\int_Q f(A + Du(x))\mathrm{d}x$$

for every cube $Q \subset \mathbb{R}^n$, any $A \in \mathcal{L}$ and every $u \in W^{1,\infty}(Q;\mathbb{R}^m)$, Q-periodic with $Du(x) \in \mathcal{L}$ for almost every x. We remark that if $\mathcal{L} = \mathbb{M}^{m\times n}$ we get the usual rank-one convexity and quasiconvexity condition and thus the new conditions generalize the classical ones.

Let $f : \mathbb{M}^{m\times n} \rightarrow \mathbb{R}$ and define $I(u) = \int_{\Omega} f(Du)\mathrm{d}x$. We say I is $\mathcal{L}$-*weakly lower semicontinuous* on $W^{1,p}$ if

$$I(u) \leq \liminf_{k\rightarrow\infty} I(u_k)$$

whenever $u_k \rightharpoonup u$ and dist$(Du_k, \mathcal{L}) \rightarrow 0$ as $k \rightarrow \infty$.

The main result we prove is that assuming the subspace $\mathcal{L}$ satisfies the *constant dimension condition* (see definition below) then $\mathcal{L}$-quasiconvexity is equivalent to the $\mathcal{L}$-weak lower semicontinuity of the functional I.

8.2 Preliminaries and Notations

Let $\mathbb{R}^n$ the usual n-dimensional Euclidean space with points $x = (x_1, x_2, \dots, x_n)$, $x_i \in \mathbb{R}$ (real numbers). Let Ω be a bounded domain in $\mathbb{R}^n$ and $Q_0 = [0,1]^n$ the unit cube in $\mathbb{R}^n$. Let $\mathbb{M}^{m \times n}$ be the set of $m \times n$ matrices. For vectors $a, b \in \mathbb{R}^n$ and matrices $\xi, \eta \in \mathbb{M}^{m \times n}$, we define the inner products by

$$a \cdot b = \sum_{j=1}^{n} a_i b_i, \quad \xi : \eta = \langle \xi, \eta \rangle = \sum_{i=1}^{m} \sum_{j=1}^{n} \xi_{ij} \eta_{ij}$$

with the corresponding Euclidean norms denoted both by $|\cdot|$. For vectors $q \in \mathbb{R}^m$, $a \in \mathbb{R}^n$, we denote by $q \otimes a$ the rank-one $m \times n$ matrix $(q_i a_j)$ and also define $0 = 0_{m \times n}$ where $0_{m \times n}$ is the $m \times n$ matrix having 0 in all entries.

A cube in $\mathbb{R}^n$ is a set

$$Q = \left\{ x \in \mathbb{R}^n \;\middle|\; x = \sum_{i=1}^{n} c_i l_i \;\middle|\; 0 \leq c_i \leq 1 \right\}$$

where $\{l_1, l_2, \ldots, l_n\}$ is an orthonormal basis of $\mathbb{R}^n$.

Denoting $\mu(\Omega)$ or $|\Omega|$ the Lebesgue measure of a measurable set Ω we have that $\mu(Q) = |Q| = 1$. A function u defined on $\mathbb{R}^n$ is called *Q-periodic* if

$$u(x) = u\left(x + \sum_{i=1}^{n} c_i l_i \right)$$

for any $x \in \mathbb{R}^n$ and any $c_i \in \mathbb{Z}$.

Let $W^{1,p}(\Omega)$ be the usual Sobolev space of scalar functions on Ω, and define $W^{1,p}(\Omega; \mathbb{R}^m)$ to be the space of vector functions $u \colon \Omega \to \mathbb{R}^m$ with each component $u^i \in W^{1,p}(\Omega)$ and we denote by Du the Jacobi matrix of u defined by

$$Du(x) = (\partial u^i / \partial x_j)_{i=1,\ldots,m}^{j=1,\ldots,n}.$$

Let $1 \leq p < \infty$. We make $W^{1,p}(\Omega; \mathbb{R}^m)$ a Banach space with the norm

$$\|u\|_{W^{1,p}(\Omega;\mathbb{R}^m)} = \left(\int_{\Omega} (|u|^p + |Du|^p) \, dx \right)^{\frac{1}{p}}$$

Let $C_0^{\infty}(\Omega; \mathbb{R}^m)$ be the set of infinitely differentiable vector functions with compact support in Ω, and let $W_0^{1,p}(\Omega; \mathbb{R}^m)$ be the closure of $C_0^{\infty}(\Omega; \mathbb{R}^m)$ in $W^{1,p}(\Omega; \mathbb{R}^m)$. Then $W_0^{1,p}(\Omega; \mathbb{R}^m)$ is itself a Banach space and has an equivalent norm defined by $\||Du|\|_{L^p(\Omega)}$. We also recall the following version of Sobolev embedding:.

Theorem 8.2.1. *If Ω is a bounded Lipschitz domain then the embedding*

$$W^{1,p}(\Omega; \mathbb{R}^m) \to L^p(\Omega; \mathbb{R}^m)$$

is compact for any $1 \leq p \leq \infty$.

By $C_0(\mathbb{R}^n)$ we denote the closure of continuous functions on $\mathbb{R}^n$ with compact support. The dual of $C_0(\mathbb{R}^n)$ can be identified with the space $\mathcal{M}(\mathbb{R}^n)$ of signed Radon measures with finite mass via the pairing

$$\langle v, f \rangle = \int_{\mathbb{R}^n} f \, dv$$

A map $v : E \to \mathcal{M}(\mathbb{R}^n)$ is called weak* measurable if the functions $x \to \langle v(x), f \rangle$ are measurable for all $f \in C_0(\mathbb{R}^n)$. We shall write v_x instead of $v(x)$.

Let $f : \Omega \times \mathbb{R}^m \to \mathbb{R}$ a function measurable in x such that $v \to f(x, v)$ is continuous for all $x \in \Omega$ (a function with this properties is called Carathéodory function). The following result represents the fundamental theorem of Young measures:.

Theorem 8.2.2 ([5]). *Let $E \subset \mathbb{R}^n$ be a measurable set of finite measure and let $u_k : E \to \mathbb{R}^m$ be a sequence of measurable functions. Then there exists a subsequence u_{k_j} and a weak* measurable map $v : E \to \mathcal{M}(\mathbb{R}^m)$ such that the following hold.*

(i) $v_x \geq 0$, $\|v_x\|_{\mathcal{M}(\mathbb{R}^m)} = \int_{\mathbb{R}^m} dv_x \leq 1$, *for almost every $x \in E$.*

(ii) *We have $\|v_x\|_{\mathcal{M}(\mathbb{R}^m)} = 1$ if and only if the sequence does not escape to infinity, i.e., if $\lim\sup\limits_{r \to \infty} |\{|u_{k_j}|\} \geq r| = 0$.*

(iii) *Let $A \subset E$ measurable and $f \in C(\mathbb{R}^m)$. If $\|v_x\|_{\mathcal{M}(\mathbb{R}^m)} = 1$ for almost every $x \in E$ and if $f(u_{k_j})$ is relatively compact in $L^1(A)$ then*

$$f(u_{k_j}) \rightharpoonup \langle v_x, f \rangle = \int_{\mathbb{R}^m} f \, dv_x$$

(iv) *If f is Carathéodory and bounded from below then*

$$\lim_{n \to \infty} \int_\Omega f(x, u_{k_j})(x)) dx = \int_\Omega \langle v_x, f(x, u_{k_j}(x)) \rangle dx < \infty$$

if and only if $\{f(\cdot, u_{k_j}(\cdot))\}$ is equi-integrable.

The measures $(v_x)_{x \in \Omega}$ are called the *Young measures* generated by the sequence $\{u_{k_j}\}$. The Young measure is said to be *homogeneous* if there is a Radon measure $v_0 \in \mathcal{M}(\mathbb{R}^m)$ such that $v_x = v_0$ for almost every $x \in \Omega$.

Theorem 8.2.3 ([34]). *If $\{u_k\}$ is a sequence of measurable functions with associated Young measure $v = \{v_x\}_{x \in \Omega}$, then*

$$\liminf_{k \to \infty} \int_E f(x, u_k(x)) dx \geq \int_E \int_{\mathbb{R}^m} f(x, \lambda) dv_x(\lambda) dx \tag{8.3}$$

for every Carathéodory function f, bounded from below, and every measurable subset $E \subset \Omega$.

A Young measure (v_x) is called a *gradient Young measure* if it is generated by a sequence of gradients. We say that (v_x) is a $W^{1,p}$ gradient Young measure if it is generated by $\{Du_k\}$ and $u_k \rightharpoonup u$ in $W^{1,p}(\Omega, \mathbb{R}^m)$. The following result refers to the localization of the gradient Young measures.

Theorem 8.2.4 ([20]). *Let* (v_x) *be a gradient Young measure generated by a sequence of gradients of functions in* $W^{1,p}(\Omega)$. *Then for almost every* $a \in \Omega$ *there exists a sequence of gradients of functions in* $W^{1,p}(\Omega)$ *that generates the homogeneous Young measure* (v_a).

We also provide the definitions of convexity, rank-one convexity, and quasiconvexity.

Definition 8.2.5. Let $h : \mathbb{M}^{m \times n} \to \mathbb{R}$. We say that h is *convex* on $\mathbb{M}^{m \times n}$ if the inequality

$$h(\lambda \xi + (1 - \lambda)\eta) \leq \lambda h(\xi) + (1 - \lambda)h(\eta) \tag{8.4}$$

holds for all $0 < \lambda < 1$ and $\xi, \eta \in \mathbb{M}^{m \times n}$.

Note also that h is convex if and only if $g(t) = h(\xi + t\eta)$ is a convex function of t on $\mathbb{R}$ for all $\xi, \eta \in \mathbb{M}^{m \times n}$. For C^1 functions h, the convexity condition is equivalent to the condition

$$h(\eta) \geq h(\xi) + D_\xi h(\xi) : (\eta - \xi), \quad \forall \eta, \xi \in \mathbb{M}^{m \times n} \tag{8.5}$$

Furthermore, a C^1 function h on $\mathbb{R}$ is convex if and only if h' is nondecreasing, or equivalently, the following condition holds:

$$(h'(a) - h'(b))(a - b) \geq 0, \quad \forall a, b \in \mathbb{R} \tag{8.6}$$

Definition 8.2.6. A function $f : \mathbb{M}^{m \times n} \to \mathbb{R}$ is called *rank one convex* if

$$f(\lambda A + (1 - \lambda)B) \leq \lambda f(A) + (1 - \lambda)f(B)$$

for all $\lambda \in [0, 1]$ and any matrices A and B such that rank $(A - B) \leq 1$.

Definition 8.2.7. A function $f : \mathbb{R}^n \to \mathbb{R}$ is called *separately convex* if $g_i(t) = f(x_1, \ldots, x_{i-1}, t, x_{i+1}, \ldots, x_n)$ is convex in t for all $1 \leq i \leq n$.

Definition 8.2.8. A function $f : \mathbb{M}^{m \times n} \to \mathbb{R}$ is said to be *quasiconvex* if

$$\int_{Q_0} f(A + Du(x))dx \geq f(A)$$

for any $A \in \mathbb{M}^{m \times n}$ and $u \in W_0^{1,\infty}(Q_0; \mathbb{R}^m)$.

If f is quasiconvex then one can show [38] that

$$f(A) = \inf_{u \in W_{\text{per}}^{1,\infty}(Q_0; \mathbb{R}^m)} \int_{Q_0} f(A + Du(x))dx$$

where $W_{\text{per}}^{1,\infty}(Q_0; \mathbb{R}^m)$ is the class of periodic functions in $W^{1,\infty}(Q_0; \mathbb{R}^m)$.

Let $\Lambda := \mathbb{Z}^N$ be the unit lattice, i.e. the additive group of points in $\mathbb{Z}^n$ with integer coordinates. We say that $f : \mathbb{R}^n \to \mathbb{R}^m$ is Λ-periodic if

$$f(x + \lambda) = f(x) \quad \text{for all } x \in \mathbb{R}^n, \lambda \in \Lambda.$$

A $\Lambda - periodic$ function f may be identified with a function f_T on the n-torus

$$T_n := \left\{ (e^{2\pi i x_1}, e^{2\pi i x_2}, \ldots, e^{2\pi i x_n}) \in \mathbb{C}^n : (x_1, x_2, \ldots, x_n) \in \mathbb{R}^n \right\}$$

through the relation

$$f_T \left(e^{2\pi i x_1}, e^{2\pi i x_2}, \ldots, e^{2\pi i x_n} \right) := f(x_1, x_2, \ldots, x_n)$$

The space $L^p(T_n)$ is identified with $L^p(Q_0)$ and $C(T_n)$ is the set of Λ-periodic continuous functions on $\bar{Q}_0$. We recall some results on Fourier transform for periodic functions. If $f \in L^1(T_n)$, then its Fourier coefficients are defined as:

$$\hat{f}(\lambda) := \int_{T_n} f(x) e^{-2\pi i x \cdot \lambda} \, dx, \quad \lambda \in \Lambda$$

Theorem 8.2.9. *We have the following:*

(i) The trigonometric polynomials

$$R(x) := \sum_{\lambda \in \Lambda'} a_\lambda e^{-2\pi i x \cdot \lambda}, \quad \Lambda' \text{ all finite subsets of } \Lambda, \ a_\lambda \in \mathbb{C}$$

are dense in $C(T_n)$ and in $L^p(T_n)$ for all $1 \le p < \infty$.
(ii) If $f \in L^2(T_n)$ then

$$f(x) = \sum_{\lambda \in \Lambda} \hat{f}(\lambda) e^{-2\pi i x \cdot \lambda}, \quad \sum_{\lambda \in \Lambda} |\hat{f}(\lambda)|^2 = \|f\|_{L^2}$$

Let $f : \Omega \times \mathbb{R}^n \times \mathbb{M}^{m \times n} \to \mathbb{R}$. We say f is *Carathéodory* if $f(x, s, \xi)$ is measurable in x for all $(s, \xi) \in \mathbb{R}^n \times \mathbb{M}^{m \times n}$ and continuous in $(s, \xi) \in \mathbb{R}^n \times \mathbb{M}^{m \times n}$ for almost every $x \in \Omega$. Define the multiple integral functional I on $W^{1,p}(\Omega; \mathbb{R}^m)$ by

$$I(u) = \int_\Omega f(x, u(x), Du(x)) \, dx, \quad u \in W^{1,p}(\Omega; \mathbb{R}^m)$$

If $f(x, s, \xi)$ is measurable in $x \in \Omega$ for all $(s, \xi) \in \mathbb{R}^n \times \mathbb{M}^{m \times n}$ and is C^1 in $(s, \xi) \in \mathbb{R}^n \times \mathbb{M}^{m \times n}$ for almost every $x \in \Omega$, we shall use the following notation to denote the derivatives of f on s and ξ:

$$D_s f(x, s, \xi) = \left(\frac{\partial f}{\partial s_1}, \ldots, \frac{\partial f}{\partial s_n} \right), \quad D_\xi f(x, s, \xi) = (\partial f / \partial \xi_{ij})_{i=1,\ldots,m}^{j=1,\ldots,n}$$

Definition 8.2.10. A functional I is said to be (sequentially) *weakly lower semi-continuous* on $W^{1,p}(\Omega; \mathbb{R}^m)$ provided

$$I(u) \leq \liminf_{k \to \infty} I(u_k) \quad \text{whenever } u_k \rightharpoonup u \text{ in } W^{1,p}(\Omega; \mathbb{R}^m). \tag{8.7}$$

The following important result has been proved by Acerbi and Fusco [2].

Theorem 8.2.11. *Assume f is Carathéodory and satisfies*

$$0 \leq f(x, s, \xi) \leq c_1 (|\xi|^p + |s|^p) + A(x),$$

where $c_1 > 0$, $p \geq 1$, and $A \in L^1(\Omega)$. Then functional I defined above is weakly lower semicontinuous on $W^{1,p}(\Omega; \mathbb{R}^m)$ if and only if $f(x, s, \cdot)$ is quasiconvex for almost every $x \in \Omega$ and all $s \in \mathbb{R}^n$; i.e., the inequality

$$f(x, s, \xi) \leq \frac{1}{|\Omega|} \int_\Omega f(x, s, \xi + D\varphi(y)) \, dy$$

holds for a.e. $x \in \Omega$, all $s \in \mathbb{R}^n$, $\xi \in \mathbb{M}^{m \times n}$ and all $\varphi \in C_0^\infty(\Omega; \mathbb{R}^m)$.

8.3   Linear Restrictions $\mathcal{L}$-Rank-One Convexity and $\mathcal{L}$-Quasiconvexity

An interesting and motivating problem is to study necessary and sufficient conditions for the weak lower semicontinuity of the operator I restricted only to a class of functions that satisfy certain linear constraints, i.e., their gradients in the sense of distributions approach a preset target linear subspace of $\mathbb{M}^{m \times n}$ by means of L^2 convergence. When the linear subspace satisfies some special condition we prove that the restricted weak lower semicontinuity is equivalent to a generalized version of quasiconvexity.

Let $\mathcal{L}$ be a linear subspace of $\mathbb{M}^{m \times n}$ and $P : \mathbb{M}^{m \times n} \to \mathbb{M}^{m \times n}$ the linear map such that $PA = 0$ if and only if $A \in \mathcal{L}$, which is actually the orthogonal projection onto the orthogonal complement of $\mathcal{L}$.

Definition 8.3.1. We say that a function $f : \mathcal{L} \to \mathbb{R}$ is $\mathcal{L}$- *rank one convex* if for any $\lambda \in [0, 1]$ and $A, B \in \mathcal{L}$ such that rank $(A - B) \leq 1$ we have

$$f(\lambda A + (1 - \lambda)B) \leq \lambda f(A) + (1 - \lambda)f(B)$$

Definition 8.3.2. Given a cube $Q \subset \mathbb{R}^n$ we say that a function $f : \mathcal{L} \to \mathbb{R}$ is $Q - \mathcal{L}$-*quasiconvex* if

$$f(A) \leq \frac{1}{|Q|} \int_Q f(A + Du(x)) \, dx$$

for any $A \in \mathcal{L}$ and every $u \in W^{1,\infty}(Q; \mathbb{R}^m)$, Q-periodic with $Du \in \mathcal{L}$.

Definition 8.3.3. We say that a function $f : \mathcal{L} \to \mathbb{R}$ is $\mathcal{L}$- *quasiconvex* if it is Q-$\mathcal{L}$-quasiconvex for every cube Q,i.e.,

$$f(A) \leq \frac{1}{|Q|} \int_Q f(A + Du(x)) dx$$

for any cube $Q \subset \mathbb{R}^n$, any $A \in \mathcal{L}$ and every $u \in W^{1,\infty}(Q; \mathbb{R}^m)$, Q-periodic with $Du \in \mathcal{L}$.

Theorem 8.3.4. *If a function $f : \mathcal{L} \to \mathbb{R}$ is $\mathcal{L}$-quasiconvex then it is also $\mathcal{L}$-rank one convex.*

Proof. Let $\lambda \in [0,1]$ and A, B two elements in the subspace $\mathcal{L}$ such that rank $(A - B) \leq 1$. Let $Q_0 = [0,1]^n$ a unit cube in $\mathbb{R}^n$. Since rank$(A - B) \leq 1$ there exist two vectors $a \in \mathbb{R}^m$ and $b \in \mathbb{R}^n$ such that

$$A - B = a \otimes b$$

and it exists a rotation, a matrix $R \in \mathbb{R}^{n \times n}$ such that $RR^{\mathrm{T}} = I_n$ and $(R^{\mathrm{T}} b)^{\mathrm{T}} = e_1$ where $e_1 \in \mathbb{R}^n$ with $e_1 = (1,0,0,\ldots,0)$. Thus $(A - B)R = a(R^{\mathrm{T}} b)^{\mathrm{T}} = a \otimes e_1$. Let $Q = RQ_0$. Since f is assumed to be $\mathcal{L}$-quasiconvex we have

$$\int_Q f(C + D\varphi(x)) dx \geq f(C) \tag{8.8}$$

for all $C \in \mathcal{L}$ and $\varphi \in W^{1,\infty}(Q, \mathbb{R}^m)$ such that is Q-periodic and $D\varphi(x) \in \mathcal{L}$ a.e. x. Let $\tilde{f}(A) = f(AR^{\mathrm{T}})$ and also denote $\tilde{A} = AR$ and $\tilde{\varphi}(x) = \varphi(Rx)$. Notice that $\tilde{\varphi}$ is Q_0-periodic,

$$D\tilde{\varphi}(x) \in \tilde{\mathcal{L}} = \{\tilde{M} | \tilde{M} = MR, M \in \mathcal{L}\}$$

almost every x and $\tilde{\varphi} \in W^{1,\infty}(Q_0, \mathbb{R}^m)$. By the change of variable under the integral we obtain

$$\int_{Q_0} \tilde{f}(\tilde{C} + D\tilde{\varphi}(x)) dx \geq \tilde{f}(\tilde{C}) \tag{8.9}$$

for all $\tilde{C} \in \tilde{\mathcal{L}}$ and all $\tilde{\varphi} \in W^{1,\infty}(Q_0, \mathbb{R}^m)$, Q_0-periodic and $D\tilde{\varphi}(x) \in \tilde{\mathcal{L}}$. Also we have

$$\tilde{A} - \tilde{B} = a \otimes e_1.$$

Let $\eta : [0,1] \to \mathbb{R}$ such that $\eta'(t) = \begin{cases} (1 - \lambda) & \text{if } t \in [0, \lambda] \\ -\lambda & \text{if } t \in [\lambda, 1] \end{cases}$ and let $\tilde{\varphi}(x) = \eta(x_1)a$

where $x = (x_1, x_2, \ldots, x_n)$. Thus we obtain that $\tilde{\varphi}$ is Q_0 periodic and we can extend by this periodicity to $\mathbb{R}^n$ and $D\tilde{\varphi}(x) \in \tilde{\mathcal{L}}$ a.e. x. Also notice that $\tilde{\varphi} \in W^{1,\infty}(\mathbb{R}^n, \mathbb{R}^m)$ and

$$D\tilde{\varphi}(x) = \begin{cases} (1 - \lambda)(\tilde{A} - \tilde{B}) & \text{if } x_1 \in [0, \lambda] \\ -\lambda(\tilde{A} - \tilde{B}) & \text{if } x_1 \in [\lambda, 1] \end{cases}$$

Thus we have that $\int\limits_{Q_0} \tilde{f}(\lambda\tilde{A} + (1-\lambda)\tilde{B} + D\tilde{\varphi}(x))dx = \lambda\tilde{f}(\tilde{A}) + (1-\lambda)\tilde{f}(\tilde{B})$ and

$\int\limits_{Q_0} \tilde{f}(\lambda\tilde{A} + (1-\lambda)\tilde{B} + D\tilde{\varphi}(x))dx \geq \tilde{f}(\lambda\tilde{A} + (1-\lambda)\tilde{B})$ and obtain $\tilde{f}(\lambda\tilde{A} + (1-$

$\lambda)\tilde{B}) \leq \lambda\tilde{f}(\tilde{A}) + (1-\lambda)\tilde{f}(\tilde{B})$ hence $f(\lambda A + (1-\lambda)B) \leq \lambda f(A) + (1-\lambda)f(B)$ □

Proposition 8.3.5. *If the subspace $\mathcal{L}$ does not contain rank one matrices and a function $u \in W^{1,2}(Q;\mathbb{R}^m)$, Q-periodic has the property that $Du(x) \in \mathcal{L}$ almost every x then $u = $ const.*

Proof. Assume first $Q=Q_0$. Since $\mathcal{L}$ does not contain rank one matrices we have that

$$\min_{|a|=1,|\lambda|=1} |P(a\otimes\lambda)| > 0 \tag{8.10}$$

and it follows that

$$|P(a\otimes\lambda)| > c|a||\lambda| \tag{8.11}$$

for any $a \in \mathbb{R}^m \setminus \{0_m\}$ and $\lambda \in \mathbb{R}^n \setminus \{0_n\}$. We consider now the Fourier transform of PDu which is $P(\hat{u}(\lambda)\otimes\lambda)$. Since $\mathcal{L}$ does not contain rank one matrices we have that

$$P(\hat{u}(\lambda)\otimes\lambda) = 0 \tag{8.12}$$

for all $\lambda \in \Lambda \setminus \{0_n\}$. Thus, using (8.11) we get that $\hat{u}(\lambda) = 0$ for all $\lambda \in \mathbb{R}^n \setminus \{0_n\}$ which proves that u must be a constant.

Now, if $Q = RQ_0$ for a rotation R and $u \in W^{1,2}(Q;\mathbb{R}^m)$, Q-periodic with $Du(x) \in \mathcal{L}$ we have that $\tilde{u}(x) = u(Rx)$ is in $W^{1,2}(Q_0;\mathbb{R}^m)$, Q_0-periodic. Also $D\tilde{u} = Du(Rx)R$ so $D\tilde{u} \in \tilde{\mathcal{L}}$ where $\tilde{\mathcal{L}} = \{\tilde{A} \in \mathbb{M}^{m\times n} | \tilde{A} = AR, A \in \mathcal{L}\}$. Since $\mathcal{L}$ doesn't contain rank one matrices it follows that $\tilde{\mathcal{L}}$ has the same property. Thus $\tilde{u}$ must be constant and therefore u is constant as well.   □

8.4 Examples

In this section we are going to discuss particular cases of linear subspaces $\mathcal{L}$ and some aspects related to the restricted rank-one convexity and quasiconvexity.

Example 8.4.1. Consider $\mathcal{L} = \left\{ \begin{pmatrix} a & b \\ b & a \end{pmatrix} | a,b \in \mathbb{R} \right\}$ and let $f : \mathcal{L} \to \mathbb{R}$ a $\mathcal{L}$-rank one convex function. We show that f must be Q_0-$\mathcal{L}$-quasiconvex.

Given $u \in W^{1,\infty}(\Omega;\mathbb{R}^m)$, $u(x,y) = (u^1(x,y), u^2(x,y))$ with $Du \in \mathcal{L}$ it implies

$$\partial_x u^1 = \partial_y u^2$$

Thus u^1 and u^2 satisfy the wave equation i.e.,

$$\partial_{xx} u^1 - \partial_{yy} u^1 = 0$$

$$\partial_{xx} u^2 - \partial_{yy} u^2 = 0$$

and we get

$$u^1(x,y) = h(x+y) - g(x-y)$$

$$u^2(x,y) = h(x+y) + g(x-y)$$

where $h, g : \mathbb{R} \to \mathbb{R}$, absolutely continuous. If u is assumed to be Q_0 periodic it follows that h and g are periodic of period 1. Indeed, $u^1(x,y) = u^1(x+1,y)$ so $h(x+y+1) - h(x+y) = g(x-y) - g(x-y+1)$ for any $x, y \in \mathbb{R}$. It implies that $g(t) - g(t+1) = g(0) - g(1)$ for any $t \in \mathbb{R}$ since if two absolutely continuous functions α and β verify $\alpha(x+y) = \beta(x-y)$ for any x, y it follows that they must be constant. Thus we get that

$$g(1) - g(k+1) = (g(0) - g(1))k$$

for any positive integer k. Since g has to be bounded, we get $g(0) - g(1) = 0$ and thus $g(t) - g(t+1) = 0$ for any $t \in \mathbb{R}$.

Let $F : \mathbb{R}^2 \to \mathbb{R}$ defined as $F(a,b) = f\binom{a+b\ \ a-b}{a-b\ \ a+b}$. Since f is $\mathcal{L}$-rank one convex, we have that F is separately convex in each variable and

$$f\left(\begin{pmatrix} c & d \\ d & c \end{pmatrix} + \begin{pmatrix} a+b & a-b \\ a-b & a+b \end{pmatrix}\right) = F\left(\frac{c+d}{2} + a, \frac{c-d}{2} + b\right) \qquad (8.13)$$

Now we prove that f is Q_0-$\mathcal{L}$-quasiconvex. Making the substitution $\xi = x+y$ and $\eta = x - y$ we get

$$\iint_{Q_0} f\left(\begin{pmatrix} c & d \\ d & c \end{pmatrix} + Du\right) dx\,dy = \frac{1}{2} \int_0^1 \left(\int_{-\xi}^{\xi} F\left(\frac{c+d}{2} + h'(\xi), \frac{c-d}{2} + g'(\eta)\right) d\eta \right) d\xi$$

$$+ \frac{1}{2} \int_1^2 \left(\int_{2-\xi}^{\xi-2} F\left(\frac{c+d}{2} + h'(\xi), \frac{c-d}{2} + g'(\eta)\right) d\eta \right) d\xi$$

Now using the fact that F is separately convex and Jenssen's inequality we get

$$\int_0^1 \xi F\left(h'(\xi), \frac{g(\xi) - g(-\xi)}{2\xi}\right) + (1-\xi)F\left(h'(\xi), \frac{g(-\xi) - g(\xi)}{2\xi}\right) d\xi$$

$$\geq \int_0^1 F(h'(\xi), 0) d\xi \geq F\left(\frac{c+d}{2}, \frac{c-d}{2}\right) = f\left(\begin{pmatrix} c & d \\ d & c \end{pmatrix}\right)$$

and it follows

$$\iint_{Q_0} f\left(\begin{pmatrix} c & d \\ d & c \end{pmatrix} + Du\right) \mathrm{dxdy} \geq f\left(\begin{pmatrix} c & d \\ d & c \end{pmatrix}\right)$$

hence f is Q_0-$\mathcal{L}$-quasiconvex.

Example 8.4.2. We show that Q_0-$\mathcal{L}$-quasiconvexity might not imply $\mathcal{L}$-rank-one convexity.

Let $\mathcal{L} = \left\{\begin{pmatrix} a & b\sqrt{2} \\ b & a\sqrt{2} \end{pmatrix} | a, b \in \mathbb{R}\right\}$ a linear subspace of $\mathbb{R}^{2 \times 2}$. If $u \in W^{1,\infty}(Q_0; \mathbb{R}^m)$ satisfies $Du \in \mathcal{L}$ it follows that

$$2\partial_{xx} u^1 - \partial_{yy} u^1 = 0 \qquad (8.14)$$

which implies that there exist $h, g : \mathbb{R} \to \mathbb{R}$ such that

$$u^1(x,y) = h\left(x + y\sqrt{2}\right) + g\left(x - y\sqrt{2}\right) \qquad (8.15)$$

Also, $u^1(x,y)$ is Q_0-periodic so we get, by reasoning as before, that h and g are periodic with periods 1 and $\sqrt{2}$. Since $\sqrt{2}$ is irrational and the set $\left\{k\sqrt{2} + p \,|\, k, p \in \mathbb{Z}\right\}$ is dense in $\mathbb{R}$ it follows that h and g must be constant. Therefore, by definition, every function is Q_0-$\mathcal{L}$-quasiconvex, but not necessarily $\mathcal{L}$-rank one convex (see Example 8.4.1).

Example 8.4.3. We show that $\mathcal{L}$-rank-one convexity does not imply $\mathcal{L}$-quasiconvexity. The following famous example belongs to Šverák [38].

Let

$$\mathcal{L} = \left\{\begin{pmatrix} a & 0 \\ 0 & b \\ c & c \end{pmatrix}, \quad a, b, c \in \mathbb{R}\right\} \qquad (8.16)$$

a linear subspace of $\mathbb{M}^{3 \times 2}$. Also let $f : \mathcal{L} \to \mathbb{R}$ be defined by

$$f\left(\begin{pmatrix} a & 0 \\ 0 & b \\ c & c \end{pmatrix}\right) = -abc \qquad (8.17)$$

We notice that the only rank-one directions in $\mathcal{L}$ are given by

$$\begin{pmatrix} 1 & 0 \\ 0 & 0 \\ 0 & 0 \end{pmatrix}, \quad \begin{pmatrix} 0 & 0 \\ 0 & 1 \\ 0 & 0 \end{pmatrix}, \text{ and } \begin{pmatrix} 0 & 0 \\ 0 & 0 \\ 1 & 1 \end{pmatrix}$$

and the function f is convex on each rank-one line contained in $\mathcal{L}$. Consider the function $u : \mathbb{R}^2 \to \mathbb{R}^3$ given by

$$u(x,y) = \frac{1}{2\pi} \begin{pmatrix} \sin(2\pi x) \\ \sin(2\pi y) \\ \sin(2\pi(x+y)) \end{pmatrix}$$

We have that $u \in W^{1,\infty}(Q_0; \mathbb{R}^3)$ where $Q_0 = [0,1]^2$, u is Q_0-periodic and $Du \in \mathcal{L}$ since

$$Du(x,y) = \begin{pmatrix} \cos(2\pi x) & 0 \\ 0 & \cos(2\pi y) \\ \cos(2\pi(x+y)) & \cos(2\pi(x+y)) \end{pmatrix}$$

Thus we get

$$\iint_{Q_0} f(Du(x,y)) dx dy = -\iint_{Q_0} (\cos(2\pi x))^2 (\cos(2\pi y))^2 dx dy < 0 = f(0_{3\times 2}) \tag{8.18}$$

which shows that f is not $\mathcal{L}$-quasiconvex.

Now we generalize Example 8.4.3 to the case where some function $f : \mathcal{L} \to \mathbb{R}$ which is $\mathcal{L}$-rank one convex but not Q_0-$\mathcal{L}$-quasiconvex can be extended to the entire space $\mathbb{M}^{m\times n}$ and preserve this property.

Theorem 8.4.4. *Let* $f : \mathcal{L} \to \mathbb{R}$ *be a function which is $\mathcal{L}$-rank one but it is not $\mathcal{L}$-quasiconvex. Also assume that f is C^2 and for some $p \geq 2$:*

$$|f(A)| \leq c(1 + |A|^p). \tag{8.19}$$

$$|D^2 f(A)| \leq c\left(1 + |A|^{p-2}\right) \tag{8.20}$$

for all $A \in \mathcal{L}$. Then there exists a function $F : \mathbb{M}^{m\times n} \to \mathbb{R}$ which is rank one convex but not quasiconvex on $\mathbb{M}^{m\times n}$.

Proof. Since f is not $\mathcal{L}$-quasiconvex it exists a cube $Q = RQ_0$ and $u \in W^{1,\infty}(Q; \mathbb{R}^m)$, Q-periodic with $Du(x) \in \mathcal{L}$ such that

$$f(0) > \int_Q f(Du(x)) dx \tag{8.21}$$

Let $F_{\varepsilon,k} : \mathbb{M}^{m\times n} \to \mathbb{R}$ with

$$F_{\varepsilon,k}(X) = f(PX) + \varepsilon|X|^2 + \varepsilon|X|^{p+1} + k|X - PX|^2. \tag{8.22}$$

Here P is the projection onto $\mathcal{L}$. Let $A, Y \in \mathbb{M}^{m\times n}$ arbitrary such that rank $Y = 1$, $|Y| = 1$ and let $h_{\varepsilon,k} = F_{\varepsilon,k}(A + tY)$. We are going to prove that for every $\varepsilon > 0$ it exists k such that $F_{\varepsilon,k}$ is $\mathcal{L}$-rank one convex. To show this it is enough to prove that $h''_{\varepsilon,k} \geq 0$.

Thus, now we prove that

$$\left.\frac{d^2}{dt^2}F_{\varepsilon,k}(A+tY)\right|_{t=0} \geq 0 \tag{8.23}$$

for any matrices $A, Y \in \mathbb{M}^{m \times n}$ with rank $Y = 1$, $|Y| = 1$. We have

$$|A+tY|^{p+1} = \left(|A+tY|^2\right)^{\frac{p+1}{2}} = \left(|A|^2 + 2t < Y, A > +t^2\right)^{p-\frac{1}{2}} \tag{8.24}$$

$$\frac{d}{dt}|A+tY|^{p+1} = (p+1)\left(|A|^2 + 2t < Y, A > +t^2\right)^{\frac{p+1}{2}} (<Y,A> +t) \tag{8.25}$$

Thus we get

$$\left.\frac{d^2}{dt^2}|A+tY|^{p+1}\right|_{t=0} = (p+1)(p-1)|A|^{p-3} < Y, A >^2 +(p+1)|A|^{p-1} \tag{8.26}$$

and

$$\left.\frac{d^2}{dt^2}F_{\varepsilon,k}(A+tY)\right|_{t=0} = \left.\frac{d^2}{dt^2}f(PA+tPY)\right|_{t=0} + 2\varepsilon + \varepsilon(p+1)|A|^{p-1}$$
$$+ \varepsilon(p+1)(p-1)|A|^{p-3} < Y, A >^2 + k|Y - PY|^2$$

Now, from (8.19), we have

$$\left.\frac{d^2}{dt^2}f(PA+tPY)\right|_{t=0} \geq -c\left(1 + |A|^{p-2}\right) \tag{8.27}$$

and

$$\left.\frac{d^2}{dt^2}F_{\varepsilon,k}(A+tY)\right|_{t=0} \geq -c\left(1 + |A|^{p-2}\right) + \varepsilon(p+1)|A|^{p-1} + 2\varepsilon + 2k|Y - PY|^2 \tag{8.28}$$

Assume by contradiction that it exists ε_0 such that for every positive integer k we get A^k, Y^k satisfying

$$0 > \left.\frac{d^2}{dt^2}F_{\varepsilon_0,k}\left(A^k + tY^k\right)\right|_{t=0} \tag{8.29}$$

From (8.28) it follows that A^k is bounded and by extracting a subsequence we have $A^k \to \bar{A}$ and $Y^k \to \bar{Y} = P\bar{Y}$ as $k \to \infty$. Thus, passing to the limit in (8.28),

$$-\varepsilon > \left.\frac{d^2}{dt^2}f(\bar{A}+t\bar{Y})\right|_{t=0} \tag{8.30}$$

a contradiction with the fact that f is $\mathcal{L}$-rank one convex.

Now we can also choose ε small such that

$$F_{\varepsilon,k}(0) > \int_Q F_{\varepsilon,k}(Du(x))dx \qquad (8.31)$$

where u is given in (2.14). Hence $F_{\varepsilon,k}$ is not $\mathcal{L}$-quasiconvex. □

8.5   The Constant Dimension Condition

Let $\lambda \in \mathbb{R}^n$ and $R_{\mathcal{L}}^\lambda = \{w \in \mathbb{R}^m | w \otimes \lambda \in \mathcal{L}\}$. We notice that $R_{\mathcal{L}}^\lambda$ is a linear subspace of $\mathbb{R}^m$.

Definition 8.5.1. We say that the subspace $\mathcal{L}$ satisfies the *constant dimension condition* if the related subspace $R_{\mathcal{L}}^\lambda$ has the same dimension for all $\lambda \in \mathbb{R}^n \setminus \{0\}$.

If $\mathcal{L}$ satisfies the constant dimension condition we shall prove the equivalence between Q_0-$\mathcal{L}$-quasiconvexity and the weak lower semicontinuity of the functional

$$I_\Omega(u) = \int_\Omega f(Du)dx$$

along sequences satisfying the linear restriction $PDu_k(x) \to 0$ almost every x.

Remark 8.5.2. If $m = n = 2$ and $\mathcal{L}$ is the linear subspace of 2×2 symmetric matrices then the dimension of $R_{\mathcal{L}}^\lambda$ is constantly 1 for all $\lambda \in \mathbb{R}^2 \setminus \{0_2\}$.

Proof. We have that $\mathcal{L} = \left\{ \begin{pmatrix} a & b \\ b & c \end{pmatrix} \Big| \ a,b,c \in \mathbb{R} \right\}$ and

$$R_L^\lambda = \left\{ w = (w_1, w_2) \in \mathbb{R}^2 | \ w_1\lambda_2 = w_2\lambda_1 \right\}$$

Clearly the dimension of $R_{\mathcal{L}}^\lambda$ is 1 for any $\lambda \in \mathbb{R}^2 \setminus \{0_2\}$ □

Lemma 8.5.3. *If $\mathcal{L}$ satisfies the constant dimension condition there exists $\gamma > 0$ such that for any $a \in (R_{\mathcal{L}}^\lambda)^\perp$ and $\lambda \in \mathbb{R}^n \setminus \{0\}$ we have*

$$|P(\lambda \otimes a)| \geq \gamma|\lambda \otimes a| \qquad (8.32)$$

Proof. Assume by contradiction that

$$\min_{|\lambda|=1, |a|=1} |P(\lambda \otimes a)| = 0 \qquad (8.33)$$

Then there exists a minimizing sequence $\lambda_j \to \bar{\lambda}$ and $a_j \to \bar{a}$. Let $k = \dim R_{\mathcal{L}}^\lambda$. For ε small enough and any λ such that $|\lambda - \bar{\lambda}| < \varepsilon$ there exists a set $w_1(\lambda), w_2(\lambda),$ $\dots, w_k(\lambda)$ of linearly independent vectors of $R_{\mathcal{L}}^\lambda$ and $\lim_{\lambda \to \bar{\lambda}} w_i(\lambda) = w_i(\bar{\lambda})$, for all i,

$1 \leq i \leq k$. Since $a_j \in (R_{\mathcal{L}}^{\lambda_j})^{\perp}$, it implies that $\langle a_j, w_i(\lambda_j) \rangle = 0$ for all i, $1 \leq i \leq k$. We get $\langle \bar{a}, w_i(\bar{\lambda}) \rangle = 0$ so $\bar{a} \in (R_{\mathcal{L}}^{\lambda})^{\perp}$. Also, since $P(\bar{\lambda} \otimes \bar{a}) = 0$ it implies $\bar{a} \in R_{\mathcal{L}}^{\lambda}$. Thus $\bar{a} = 0$, in contradiction with $|\bar{a}| = 1$.   □

First we shall prove the selection theorem.

Theorem 8.5.4. *Let Q a cube in $\mathbb{R}^n$ and $u \in W^{1,2}(Q; \mathbb{R}^m)$ a Q-periodic function. If the linear subspace $\mathcal{L}$ satisfies the constant dimension condition then for every $\varepsilon > 0$ there exists a selection v_{ε}, $v_{\varepsilon} \in C^{\infty}$ a Q-periodic function such that $Dv_{\varepsilon}(x) \in \mathcal{L}$ a.e. $x \in Q$ and*

$$\|Du - Dv_{\varepsilon}\|_{L^2(Q)} \leq \|PDu\|_{L^2(Q)} + \varepsilon \qquad (8.34)$$

Proof. First we assume that $Q = Q_0$. Let $\Lambda = \mathbb{Z}^n$ be the unit lattice, i.e., the additive group of points in $\mathbb{R}^n$ with integer coordinates. Since u is Q-periodic we can expand u as a Fourier series

$$u(x) = \sum_{\lambda \in \Lambda} \hat{u}(\lambda) e^{2\pi i \lambda x}$$

Thus $Du(x) = \sum_{\lambda \in \Lambda} \hat{u}(\lambda) \otimes \lambda e^{2\pi i \lambda x}$. Let $\hat{v}(\lambda) = \mathbb{P}_{R_{\mathcal{L}}^{\lambda}} \hat{u}(\lambda)$, projection of both real part and imaginary part of $\hat{u}(\lambda)$ onto $R_{\mathcal{L}}^{\lambda}$. By Riesz–Fischer theorem we have that

$$v(x) = \sum_{\lambda \in \Lambda} \hat{v}(\lambda) e^{2\pi i \lambda x} \qquad (8.35)$$

is a function in $W^{1,2}(Q)$, Q-periodic and its gradient belong to $\mathcal{L}$ almost every x. Applying Lemma 8.5.3 for $a = \hat{u}(\lambda) - \hat{v}(\lambda)$ we get $\|Du - Dv\|_2 \leq \|PDu\|_2$.

Now we can consider $v_{\varepsilon}(x)$ as the real part of $\sum_{\lambda \in \Lambda'} \hat{v}(\lambda) e^{2\pi i \lambda x}$ where is Λ' is a finite subset of Λ such that

$$\|Dv_{\varepsilon} - Dv\|_{L^2} < \varepsilon \qquad (8.36)$$

since the imaginary part of $\sum_{\lambda \in \Lambda'} \hat{v}(\lambda) \otimes \lambda e^{2\pi i \lambda x}$ converges to $0_{m \times n}$ as $\Lambda' \nearrow \Lambda$.

Now if the cube Q is arbitrary then $Q = SQ_0$ for some $a \in \mathbb{R}^n$ and a rotation S. Let $\tilde{\mathcal{L}} = \{\tilde{A} \in \mathbb{M}^{m \times n} | \tilde{A} = AS, A \in \mathcal{L}\}$ and $\tilde{P}$ the orthogonal projection onto $\tilde{\mathcal{L}}$. Define $\tilde{u} : Q_0 \to \mathbb{R}^m$ by $\tilde{u}(x) := u(Sx)$. Also notice that

$$R_{\tilde{\mathcal{L}}}^{\lambda} = R_{\mathcal{L}}^{S\lambda}$$

and therefore $R_{\tilde{\mathcal{L}}}^{\lambda}$ has constant dimension for any $\lambda \in \mathbb{R}^n$. Thus we can select $\tilde{v}$ such that $\tilde{v} \in C^{\infty}(Q_0)$, Q_0-periodic and

$$\|D\tilde{u} - D\tilde{v}_{\varepsilon}\|_{L^2(Q_0)} \leq \|\tilde{P}D\tilde{u}\|_{L^2(Q_0)} + \varepsilon \qquad (8.37)$$

For each $x \in Q$ there exists a unique $\bar{x} \in Q_0$ such that $x = S\bar{x}$. Let $v : Q \to \mathbb{R}^m$ with $v_{\varepsilon}(x) = \tilde{v}_{\varepsilon}(S^T(x))$. We notice that v_{ε} satisfies the requirement of the lemma.   □

8.6 $\mathcal{L}$-Weak Lower Semicontinuity

Let $f : \mathbb{M}^{m \times n} \to \mathbb{R}$ satisfy the growth condition

$$|f(A)| \le c\left(1 + |A|^2\right) \tag{8.38}$$

for any matrix $A \in \mathbb{M}^{m \times n}$ and consider the integral operator

$$I_\Omega(u) = \int_\Omega f(Du)\mathrm{d}x \tag{8.39}$$

where Ω is open bounded domain with Lipschitz boundary and $u \in W^{1,2}(\Omega; \mathbb{R}^m)$.

In contrast to Example 8.4.2 in Sect. 2.2 we show that under the constant dimension condition Q_0-$\mathcal{L}$-quasiconvexity implies $\mathcal{L}$-rank one convexity.

Theorem 8.6.1. *Assume that the linear subspace $\mathcal{L}$ satisfies the constant dimension condition. If a continuous function $f : \mathbb{M}^{m \times n} \to \mathbb{R}$ satisfies the growth condition (8.38) and is Q_0-$\mathcal{L}$-quasiconvex then it is also $\mathcal{L}$-rank one convex.*

Proof. Let $A, B \in \mathcal{L}$ be such that $\mathrm{rank}(A - B) \le 1$ and $\lambda \in [0, 1]$. For any integer k there exist $Q_1^k, Q_2^k \subset Q_0$, $Q_1^k \cap Q_2^k = \emptyset$ and $\varphi_k \in W_0^{1,\infty}(Q_0, \mathbb{R}^m)$ such that

$$\begin{cases} |\mu(Q_1^k) - \lambda| \le \frac{1}{k} \\ |\mu(Q_2^k) - (1 - \lambda)| \le \frac{1}{k} \\ D\varphi_k(x) = \begin{cases} (1 - \lambda)(A - B) & \text{if } x \in Q_1^k \\ -\lambda(A - B) & \text{if } x \in Q_2^k \end{cases} \\ \|D\varphi_k\|_\infty \le \mathrm{const}(A, B) \end{cases}$$

since $\mu(Q_0) = 1$ (see [12]). We extend the φ_k to be Q_0-periodic on $\mathbb{R}^n$. From these properties we also have that $PD\varphi_k \to 0$ in $L^2(Q_0)$. Thus, by Theorem 8.5.4, for any ε we can find a selection $u_{k,\varepsilon} \in W^{1,\infty}(Q_0, \mathbb{R}^m)$, Q_0-periodic such that $Du_{k,\varepsilon} \in \mathcal{L}$ and

$$\|Du_{k,\varepsilon} - D\varphi_k\|_{L^2(Q_0)} \to 0$$

as $\varepsilon \to 0$ and it follows

$$\liminf_{k \to \infty, \varepsilon \to 0} \int_{Q_0} f(\lambda A + (1 - \lambda B) + Du_{k,\varepsilon})\mathrm{d}x = \liminf_{k \to \infty} \int_{Q_0} f(\lambda A + (1 - \lambda B) + D\varphi_k)\mathrm{d}x$$

$$= \lambda f(A) + (1 - \lambda)f(B)$$

Since f is Q_0-$\mathcal{L}$-quasiconvex we have

$$\int_{Q_0} f(\lambda A + (1 - \lambda B) + Du_{k,\varepsilon})\mathrm{d}x \ge f(\lambda A + (1 - \lambda B)) \tag{8.40}$$

for any k and ε. Taking $\liminf$ over k and ε for the left-hand side of the previous inequality we obtain

$$\lambda f(A) + (1-\lambda)f(B) \geq f(\lambda A + (1-\lambda B))$$

which proves that f is $\mathcal{L}$-rank one convex. $\square$

Definition 8.6.2. Let f and I_Ω be defined as above. We say that the functional I_Ω is $\mathcal{L}$- *weakly lower semicontinuous* on $W^{1,2}(\Omega;\mathbb{R}^m)$ if for any sequence $u_k \rightharpoonup u$ in $W^{1,2}(\Omega;\mathbb{R}^m)$ with $\|PDu_k\|_{L^2(\Omega)} \to 0$ as $k \to \infty$, we have

$$I_\Omega(u) \leq \liminf_{k\to\infty} I_\Omega(u_k) \tag{8.41}$$

Theorem 8.6.3. *If the functional I_Ω is $\mathcal{L}$-weakly lower semicontinuous then the functional f is $\mathcal{L}$-quasiconvex.*

Proof. Let $Q = RQ_0, A \in \mathcal{L}$ arbitrary and $u \in W^{1,\infty}(Q;\mathbb{R}^m)$, Q-periodic with $Du(x) \in \mathcal{L}$ for almost every x. We show that

$$\int_Q f(A + Du(x))\mathrm{d}x \geq f(A) \tag{8.42}$$

assuming that I is $\mathcal{L}$-weakly lower semicontinuous. For any test function φ we have

$$\int_Q Du(kx)\varphi(x)\mathrm{d}x = \int_{Q_0} Du(kR\tilde{x})\varphi(\tilde{x})\mathrm{d}\tilde{x}$$

Thus, by Riemann–Lebesgue theorem, we have that

$$\lim_{k\to\infty} \int_Q Du(kx)\varphi(x)\mathrm{d}x = \int_{Q_0} Du(R\tilde{x})\varphi(\tilde{x})\mathrm{d}\tilde{x} = \int_Q Du(x)\mathrm{d}x \tag{8.43}$$

Let $u_k(x) = \frac{1}{k}u(kx) + Ax$. We notice that $Du_k(x) = Du(kx) + A$ and $Du_k(x) \in \mathcal{L}$ for any k and almost every x. We have that

$$Du_k \overset{*}{\rightharpoonup} A \tag{8.44}$$

and also

$$\int_Q f(A + Du(x))\mathrm{d}x = k^n \int_{\frac{1}{k}Q} f(A + Du(kx))\mathrm{d}x \tag{8.45}$$

For k sufficiently large there exist p_k cubes, $Q_1, Q_2, \ldots, Q_{p_k}$, which are translates of $\frac{1}{k}Q$ by multiples of $\frac{1}{k}$, mutually disjoint, such that

$$\bigcup_{i=1}^{p_k} Q_i \subset \Omega \text{ and } \mu\left(\Omega \setminus \bigcup_{i=1}^{p_k} Q_i\right) < \varepsilon_k \tag{8.46}$$

where $\varepsilon_k \to 0$ as $k \to \infty$. Thus we also get that $\frac{p_k}{k^n} \to \mu(\Omega)$ as $k \to \infty$.
Since I is $\mathcal{L}$-weakly lower semicontinuous it follows:

$$\liminf_{k\to\infty} \int_\Omega f(Du_k(x))dx \geq f(A)\mu(\Omega) \tag{8.47}$$

Also, from (8.45) we get

$$\int_\Omega f(Du_k(x))dx = p_k \int_{\frac{1}{k}Q} f(A+Du(kx))dx + \int_{\Omega\setminus\bigcup_{i=1}^{p_k} Q_i} f(A+Du(kx))dx$$

$$= \frac{p_k}{k^n} \int_Q f(A+Du(x))dx + \varepsilon_k C$$

Letting $k \to \infty$ we have $\mu(\Omega) \int_Q f(A+Du(x))dx \geq f(A)\mu(\Omega)$ and after dividing by $\mu(\Omega)$ we obtain what we had to prove.   □

Next we show under the constant dimension condition the $\mathcal{L}$-quasiconvexity is always sufficient for the $\mathcal{L}$-weak lower semicontinuity.

Theorem 8.6.4. *If the linear subspace $\mathcal{L}$ satisfies the constant dimension condition and if the function f is bounded from below, satisfies the growth condition (8.38), and is Q_0-$\mathcal{L}$-quasiconvex then functional I_Ω is $\mathcal{L}$-weakly lower semicontinuous on $W^{1,2}(\Omega; \mathbb{R}^m)$.*

Proof. Let $u_k \in W^{1,2}(\Omega, \mathbb{R}^m)$ such that $u_k \rightharpoonup u$ in $W^{1,2}$ and $PDu_k \to 0$ in L^2. We assume that for almost every x, Du_k generates a parametrized Young measure $(\nu_x)_{x\in\Omega}$. Then

$$Du(x) = \int_{\mathbb{M}^{m\times n}} \lambda d\nu_x(\lambda)$$

By Theorem 8.2.3 we also have that

$$\liminf_k \int_\Omega f(Du_k(x))dx \geq \int_\Omega \int_{\mathbb{M}^{m\times n}} f(\lambda)d\nu_x(\lambda)dx \tag{8.48}$$

For our purpose it would be sufficient to show

$$\int_\Omega \int_{\mathbb{M}^{m\times n}} f(\lambda)d\nu_x(\lambda)dx \geq \int_\Omega f(Du(x))dx \tag{8.49}$$

Now we actually prove

$$\int_{\mathbb{M}^{m\times n}} f(\lambda)d\nu_a(\lambda) \geq f\left(\int_{\mathbb{M}^{m\times n}} \lambda d\nu_a(\lambda)\right) = f(Du(a)) \tag{8.50}$$

for almost every $a \in \Omega$.

By Theorem 8.2.3 we have that v_a is also a gradient Young measure for almost every $a \in \Omega$. Consider a cube $Q \subset \Omega$ such that $a \in Q$. There exists $w_k \in W^{1,2}(Q)$ such that Dw_k generates v_a and $w_k \to \bar{w}$ in L^2, by the Sobolev embedding. Also we get that $Dw_k \rightharpoonup Du(a) = D\bar{w}$ and by the fundamental theorem of Young measures $PDw_k \to 0$ in $L^2(Q)$.

Let $\varphi_j \in C_0^\infty(Q)$ such that $\varphi_j \nearrow 1$ and $v_{k,j} = \varphi_j(w_k - \bar{w})$. Since $w_k \to \bar{w}$ in L^2 for each j there exists k_j such that

$$\|D\varphi_j \otimes (w_{k_j} - \bar{w})\|_{L^2(Q)} < \frac{1}{j}$$

Thus we can select a subsequence of $v_{k,j}$ which we can conveniently denote by v_k and we have $v_k \in W_0^{1,2}(Q)$ and

$$\|Dv_k - D(w_k - \bar{w})\|_{L^2(Q)} \to 0 \tag{8.51}$$

By using Theorem 8.5.4, we can select $\tilde{v}_k \in C^\infty(Q)$, Q-periodic such that $\|D\tilde{v}_k - Dv_k\|_{L^2(Q)} \to 0$ in $L^2(Q)$ and $D\tilde{v}_k(x) \in \mathcal{L}$ almost every x. So we have

$$\liminf_k \int_Q f(Du(a) + D(w_k(x) - \bar{w}(x))) dx = \liminf_k \int_Q f(Du(a) + D\tilde{v}_k(x)) dx$$

Also since f is $\mathcal{L}$-quasiconvex

$$\int_Q f(Du(a) + D\tilde{v}_k(x)) \geq f(Du(a)) dx$$

Thus it follows that

$$\liminf_k \int_\Omega f(Du(a) + D(w_k(x) - \bar{w}(x))) dx = \int_{\mathbb{M}^{m \times n}} f(\lambda) dv_a(\lambda) \geq f(Du(a))$$

This completes the proof. □

8.7 Particular Case Without the Constant Dimension Condition

Consider the linear subspace $\mathcal{L} = \left\{ \begin{pmatrix} a & 0 \\ 0 & b \end{pmatrix} \mid a, b \in \mathbb{R} \right\}$. We notice that the subspace

$$R_\mathcal{L}^\lambda = \{ w \in \mathbb{R}^2 \mid w \otimes \lambda \in \mathcal{L} \}$$

does not have constant dimension for all $\lambda \in \mathbb{R}^2 \setminus \{0\}$. Therefore this space $\mathcal{L}$ does not satisfy the constant dimension condition defined above.

Let $f : \mathbb{M}^{2 \times 2} \to \mathbb{R}$ be a C^1 function satisfying

$$0 \leq f(\xi) \leq c\left(1 + |\xi|^2\right) \tag{8.52}$$

$$|Df(\xi)| \leq c(1 + |\xi|) \tag{8.53}$$

Also, as above, define

$$I_\Omega(u) = \int_\Omega f(Du)\mathrm{d}x \tag{8.54}$$

Theorem 8.7.1. *If $f : \mathbb{M}^{2 \times 2} \to \mathbb{R}$ satisfies (8.52) and (8.53) and is $\mathcal{L}$-rank one convex then I_Ω is $\mathcal{L}$-weakly lower semicontinuous on $W^{1,2}(\Omega; \mathbb{R}^2)$.*

The following result by Müller is going to be essential in the course of the proof.

Theorem 8.7.2 ([30]). *Let $f : \mathbb{R}^2 \to \mathbb{R}$ be a separately convex function that satisfies*

$$0 \leq f(\xi) \leq C\left(1 + |\xi|^2\right)$$

Let $\Omega \subset \mathbb{R}^2$ be open and suppose that

$$u_k \rightharpoonup u, \quad v_k \rightharpoonup v \text{ in } L^2_{\mathrm{loc}}(\Omega) \tag{8.55}$$

$$\partial_y u_k \to \partial_y u, \quad \partial_x v_k \to \partial_x v, \text{ in } H^{-1}_{\mathrm{loc}}(\Omega) \tag{8.56}$$

Then we have

$$\liminf_{k \to \infty} \int_\Omega f(u_k, v_k)\mathrm{d}z \geq \int_\Omega f(u, v)\mathrm{d}z \tag{8.57}$$

Now we are going to prove Theorem 8.7.1.

Proof. Let $u_k \in W^{1,2}(\Omega; \mathbb{R}^2)$ with $u_k \rightharpoonup u$ and $PDu_k \to 0$ almost everywhere. Thus we have that $\partial_y u_k^1 \to 0$ and $\partial_x u_k^2 \to 0$ so $\partial_x(\partial_y u_k^1) \to 0$ and $\partial_y(\partial_x u_k^2) \to 0$ in $H^{-1}(\Omega)$.

Let $F : \mathcal{L} \to \mathbb{R}$ given by $F(a, b) = f\left(\begin{pmatrix} a & 0 \\ 0 & b \end{pmatrix}\right)$. Since f is $\mathcal{L}$-rank one convex it follows that F is separately convex and satisfies the growth condition from Theorem 8.7.2. From (8.52) and (8.53) we also have that

$$|f(\xi) - f(\eta)| \leq c(1 + |\xi| + |\eta|)(\xi - \eta) \tag{8.58}$$

By using this inequality with $\xi = Du_k$ and $\eta = \begin{pmatrix} \partial_x u_k^1 & 0 \\ 0 & \partial_y u_k^2 \end{pmatrix}$ we get

$$\liminf_{k \to \infty} \int_\Omega f(Du_k)\mathrm{d}z = \liminf_{k \to \infty} \int_\Omega f\left(\begin{pmatrix} \partial_x u_k^1 & 0 \\ 0 & \partial_y u_k^2 \end{pmatrix}\right)\mathrm{d}z \tag{8.59}$$

From Theorem 8.7.2 we obtain

$$F(\partial_x u^1, \partial_y u^2) \leq \liminf_{k \to \infty} \int_\Omega F\left(\partial_x u_k^1, \partial_y u_k^2\right) dz \qquad (8.60)$$

and since $\partial_y u^1 = 0$ and $\partial_x u^2 = 0$ we finally get

$$f(Du) \leq \liminf_{k \to \infty} \int_\Omega f(Du_k) dz \qquad \square$$

Remark 8.7.3. From Theorems 8.7.1 and 8.7.2 we get that for this particular case of subspace $\mathcal{L}$, every $\mathcal{L}$-rank one convex function is $\mathcal{L}$-quasiconvex but it may very difficult to prove this directly.

References

1. Adams, R.: Sobolev Spaces. Academic Press (1975)
2. Acerbi, E., Fusco, N.: Semicontinuity problems in the calculus of variations. Arch. Rational Mech. Anal. **86**, 125–145 (1984)
3. Aubin, J.-P., Ekeland, I.: Applied Nonlinear Analysis. Wiley, New York (1984)
4. Ball, J.M.: Convexity conditions and existence theorems in nonlinear elasticity. Arch. Rational Mech. Anal. **63**, 337–403 (1977)
5. Ball, J.M.: A version of the fundamental theorem for Young measures. In: Partial Differential Equations and Continuum Models of Phase Transitions. Lecture Notes in Physics, vol. 344. Springer, Berlin (1988)
6. Ball, J.M., James, R.D.: Proposed experimental tests of a theory of fine microstructures and the two well problem. Phil. Trans. Roy. Soc. London. **338A**, 389–450 (1992)
7. Ball, J.M., Murat, F.: $W^{1,p}$-Quasiconvexity and variational problems for multiple integrals. J. Funct. Anal. **58**, 225–253 (1984)
8. Brooks, J., Chacon, R.: Continuity and compactness of measures. Adv. Math. **37**, 16–26 (1980)
9. Buttazzo, G.: Semicontinuity, Relaxation and Integral Representation in the Calculus of Variations. Pitman Research Notes in Mathematics, Vol. 207. Longman, Harlow (1989)
10. Cellina, A.: On minima of a functional of the gradient: Sufficient conditions. Nonlinear Anal. **20**(4), 343–347 (1993)
11. Coifman, R., Lions, P.-L., Meyer, Y., Semmes, S.: Compensated compactness and Hardy spaces. J. Math. Pures Appl. **72**(9), 247–286 (1993)
12. Dacorogna, B.: Direct Methods in the Calculus of Variations. Springer, New York (1989)
13. Dacorogna, B., Marcellini, P.: 'Implicit Partial Differential Equations. Birkhäuser, Boston (1999)
14. De Figueiredo, D.G.: The Ekeland Variational Principle with Applications and Detours. Tata Institute Lecture, Springer, Berlin (1989)
15. Ekeland, I.: On the variational principle. J. Math. Anal. Appl. **47**, 324–353 (1974)
16. Evans, L.C.: Quasiconvexity and partial regularity in the calculus of variations Arch. Rational Mech. Anal. **95**, 227–252 (1986)
17. Evans, L.C.: Weak Convergence Methods for Nonlinear Partial Differential Equations. In: CBMS Regional Conference Series in Mathematics, Vol. 74. AMS, Providence, RI (1990)
18. Fonseca, I., Müller, S.: A-quasiconvexity, lower semicontinuity and Young measures. SIAM J.Math. Anal. **30**, 1355–1390 (1999)

19. Giaquinta, M.: Multiple Integrals in the Calculus of Variations and Nonlinear Elliptic Systems. Princeton University Press, Princeton (1983)
20. Kinderlehrer, D., Pedregal, P.: Gradient Young measures generated by sequences in sobolev spaces. J. Geomet. Anal. **4**(1) (1994)
21. Lee, J., Müller, P.F.X., Müller, S.: Compensated Compactness, Separately Convex Functions and Interpolatory Estimates between Riesz Transforms and Haar Projections. Preprint Max Planck Institute (2008)
22. Marcellini, P.: On the definition and the lower semicontinuity of certain quasiconvex integrals. Ann. Inst. H.Poincaré Analyse non linéaire. **3**(5), 391–409 (1986)
23. Marcellini, P., Sbordone, C.: On the existence of minima of multiple integrals. J. Math. Pures Appl. **62**, 1–9 (1983)
24. Meyers, N., Elcrat, A.: Some results on regularity for solutions of non-linear elliptic systems and quasi-regular functions. Duke Math. J. **42**, 121–136 (1975)
25. Milnor, J.: Topology from the Differentiable Viewpoint. University of Virginia Press (1965)
26. Morrey, C.B.: Quasiconvexity and the lower semicontinuity of multiple integrals. Pacific J. Math. **2**, 25–53 (1952)
27. Morrey, C.B.: Multiple Integrals in the Calculus of Variations. Springer, Berlin (1966)
28. Müller, S.: Higher integrability of determinants and weak convergence in L^1,. J. Reine Angew. Math. **412**, 20–34 (1990)
29. Müller, S.: Variational models for microstructure and phase transitions. In: Calculus of Variations and Geometric Evolution Problems, (Cetraro, 1996). Lecture Notes in Math, Vol. 1713, pp 85–210. Springer (1999)
30. Müller, S.: Rank-one convexity implies quasiconvexity on diagonal matrices. Int. Math. Res. Not. **20**, 1087–1095 (1999)
31. Müller, S., Šverák, V.: Attainment results for the two-well problem by convex integration. In: Jost, J. (ed.) Geometric Analysis and the Calculus of Variations, pp 239–251. Internat. Press, Cambridge, MA (1996)
32. Murat, F.: A survey on compensated compactness. In: Contributions to modern calculus of variations (Bologna, 1985), 145–183, Putman Res. Notes Math. Ser., 148, Longman Sci. Tech., Harlow (1987)
33. Palombaro, M., Smyshlyaev, V.P.: Relaxation of three solenoidal wells and characterization of extremal three-phase H-measures. Arch. Ration. Mech. Anal. **194**(3), 775–722 (2009)
34. Pedregal, P.: Parameterized Measures and Variational Principles. Birkhäuser, Besel (1997)
35. Rabinowitz, P.H.: Minimax Methods in Critical Point Theory with Applications to Differential Equations. In: CBMS Regional Conference Series in Mathematics, Vol. 65. AMS, Providence, RI (1986)
36. Santos, P.: $\mathcal{A}$-quasiconvexity with variable coefficients. Proc. Roy. Soc. Edinburgh. **134**(6), 1219–1237(19) (2004)
37. Stein, E.: Singular Integrals and Differentiability Properties of Functions. Princeton University Press, Princeton (1970)
38. Šverák, V.: Rank-one convexity does not imply quasiconvexity. Proc. Roy. Soc. Edinburgh. **120A**, 185–189 (1992)
39. Šverák, V.: On the problem of two wells. In: Kinderlehrer, D. et al. (eds) Microstructure and Phase Transition, IMA Math. Appl. Vol. **54**, pp 183–190. Springer, New York (1993)
40. Šverák, V.: Lower semicontinuity for variational integral functionals and compensated compactness. In: Chatterji, S.D. (ed.) Proceedings of the International Congress of Mathematicians, Zürich, 1994, pp 1153–1158. Birkhäuser, Basel (1995)
41. Rockafellar, R.T.: Convex Analysis. Princeton University Press, Princeton, NJ (1972)
42. Tartar, L.: Compensated compactness and applications to partial differential equations. In: Nonlinear Analysis and Mechanics: Heriot-Watt Symposium, Vol. IV, pp 136–212. Res. Notes in Math. Vol. 39. Pitman, Boston, Mass.-London (1979)
43. Yan, B.: Remarks about $W^{1,p}$-stability of the conformal set in higher dimensions. Ann. Inst. H. Poincaré, Analyse non linéaire. **13**(6), 691–705 (1996)

44. Yan, B.: On rank-one convex and polyconvex conformal energy functions with slow growth. Proc. Roy. Soc. Edinburgh. **127A**, 651–663 (1997)
45. Yan, B., Zhou, Z.: A theorem on improving regularity of minimizing sequences by reverse Hölder inequalities. Michigan Math. J. **44** (1997), 543–553.
46. Yan, B., Zhou, Z.: Stability of weakly almost conformal mappings. Proc. Am. Math. Soc. **126**, 481–489 (1998)
47. Young, L.C.: Lectures on Calculus of Variations and Optimal Control Theory. W.B. Saunders (1969)
48. Zhang, K.: A construction of quasiconvex functions with linear growth at infinity. Ann. Scuola Norm. Sup. Pisa. **19**, 313–326 (1992)
49. Zhang, K.: On various semiconvex hulls in the calculus of variations. Calc. Var. Partial Differ. Equat. **6**, 143–160 (1998)

Chapter 9
Tool Support for Efficient Programming of Graphics Processing Units

Kostadin Damevski

Abstract Graphics Processing Units (GPU) have established themselves as effective platforms for high-performance computing. Utilizing the power of these devices usually requires significant changes to existing codes or the development of a completely new solution. In this paper, we survey approaches that we believe are the most promising in reducing the complexity of programming or porting codes to GPUs. We also focus our presentation on our refactoring tool developed for this purpose, called ExtractKernel, which transforms existing C loops into code that can execute on the GPU.

Keywords Programming • Graphics Processing Units

9.1 Introduction

Graphics Processing Units (GPUs) are an emerging hardware platform for achieving high performance in a large number of applications. GPUs outperform conventional CPUs by one or two orders of magnitude in many computationally intensive tasks, due to their large number of simple computational cores (448 cores in NVIDIA's Tesla C2070 GPU) and lower memory latency. These devices are programmed in a data parallel fashion and require the presence of a CPU to manage memory allocation and transfer. Since CPUs continue to achieve higher performance than GPUs for serial tasks, CPUs and GPUs are commonly coupled as a hybrid computational platform that represents the cutting edge in high-performance computing.

K. Damevski (✉)
Department of Mathematics and Computer Science, Virginia State University,
1 Hayden Dr., Petersburg, VA 23806, USA
e-mail: kdamevski@vsu.edu

B. Toni et al. (eds.), *Bridging Mathematics, Statistics, Engineering and Technology,*
Springer Proceedings in Mathematics & Statistics 24, DOI 10.1007/978-1-4614-4559-3_9,
© Springer Science+Business Media New York 2012

A number of novel uses have recently been proposed for such hybrid computing platforms, addressing problems in virus protection, networking, image processing and scientific computing [4, 5, 7]. New programming models targeting GPUs, such as NVIDIA's Compute Unified Device Architecture (CUDA) and the Kronos Group's Open Compute Language (OpenCL), closely followed the emergence of these devices and are part of the reason for their success. However, at present, these programming models greatly lag in programmer productivity compared to mainstream programming approaches, forcing programmers to be concerned with a number of low level details, such as memory allocation and transfer between CPU and GPU.

In the scientific and high-performance computing domains, porting to new platforms, such as GPUs, has been shown to be one of the largest barriers to high programmer productivity. Several studies of programmers cite the difficulty and time overhead in porting scientific code to the latest class of supercomputers [1,3,9], time that could have been spent in developing new functionality and speeding the path to new scientific discoveries in a number of disciplines. In this paper, we examine the software engineering tool support available for constructing efficient GPU code, while reducing the programming time invested in this task. We survey a number of existing approaches, including a code refactoring approach, which was introduced by this paper's author and his collaborators.

Section 9.2 provides an overview of the task of programming GPUs and an intuition of the task's complexity. A survey of the tool support for programming GPUs is given in Sect. 9.3, including a description of the EXTRACT KERNEL refactoring, which was created by this paper's author. We limit our discussion only to programming tools and purposefully leave out discussion of tools that support other related tasks, such as debugging, testing, and profiling. Section 9.4 concludes the paper.

9.2 Programming Graphics Processing Units

In this section, we outline the complexity of writing GPU code and in porting existing code to use these devices. To show this, we use a simple example that computes the AXPY (Alpha X Plus Y), $\alpha x + y$, operation on two vectors of equivalent size N, x, and y (see Fig. 9.1). The example is written in NVIDIA's CUDA, which is currently the most common way of programming GPUs, and is also relatively similar to its alternative, OpenCL.

CUDA introduces a minor extension of the C language and a set of libraries exposed through conventional API calls. To a CUDA programmer, a program consists of two parts: one that executes on the CPU (or host) and one that executes on the GPU (or device). The GPU part of the code consists mainly of data-parallel functions, called *kernels*. To use the GPU, CUDA code follows the following workflow:

Fig. 9.1 AXPY serial code

```
 1 #define N 1000
 2
 3 int main()
 4 {
 5   float x[N], y[N], a;
 6
 7   for(int i=0; i < N; i++)
 8   {
 9     y[i] = a*x[i] + y[i];
10   }
11 }
```

1. Allocate and copy necessary data into the GPU memory (host).
2. Specify the number of threads and launch the kernel (host).
3. Execute the kernel function, placing the results in globally accessible memory (device).
4. Copy results back to the CPU memory (host).
5. Free memory on the GPU (host).

Figure 9.2 contains the CUDA code for the AXPY operation. Lines 3–10 contain the CUDA parallel kernel, which is launched with one thread per vector element. The transfer of data to the GPU memory is performed by lines 16–23, and the inverse operation is in lines 29–34 where this memory is also deallocated. In CUDA, threads are organized in thread blocks, and lines 25 and 26 calculate the number of blocks based on the assumption that the architecture supports 256 threads per block. Line 27 invokes the gpu_axpy CUDA kernel, passing pointers to the data in GPU memory and specifying the number of threads and blocks. The code to determine whether there was an error in the kernel execution is omitted here for simplicity. The memory transfer functions: cudaMalloc, cudaMemcpy, and cudaFree follow the structure and semantics of similar functions in the standard C library.

Within the CUDA kernel code, lines 6 and 7 calculate a unique thread identifier using several built-in variables that were set by the kernel invocation: blockIdx, blockDim, and threadIdx. These correspond to the index of the thread block, the number of thread blocks, and the thread index within the block. Line 9 performs the AXPY calculation, where each thread performs the multiplication and addition for the array element corresponding to its thread identifier.

9.3 Tool Support for Programming GPUs

Matlab is a programming and rapid prototyping environment that has a broad scientific user community, which is capable, in a limited way, of using GPUs to accelerate computation. Matlab's use of GPUs comes in two varieties: (1) via a small set of predefined Matlab native functions that can execute on special GPU arrays and (2) by providing the means to call CUDA functions directly from Matlab.

Fig. 9.2 AXPY kernel in
CUDA.

```
1  #define N 1000
2
3  __global__ void gpu_axpy(float a, float *x,
4                           float *y)
5  {
6    int idx = blockIdx.x * blockDim.x +
7              threadIdx.x;
8    if(idx < N)
9      y[idx] = a * x[idx] + y[idx];
10 }
11
12 int main()
13 {
14   float x[N], y[N], a;
15
16   float *x_d;
17   cudaMalloc((void **) &x_d, sizeof(float)*N);
18   cudaMemcpy(x_d, x, sizeof(float)*N,
19              cudaMemcpyHostToDevice);
20   float *y_d;
21   cudaMalloc((void **) &y_d, sizeof(float)*N);
22   cudaMemcpy(y_d, y, sizeof(float)*N,
23              cudaMemcpyHostToDevice);
24
25   int nThreads = 256;
26   int nBlocks = N / numThreads + 1;
27   gpu_axpy<<<nBlocks,nThreads>>>(a, x_d, y_d);
28
29   cudaMemcpy(x, x_d, sizeof(float)*N,
30              cudaMemcpyDeviceToHost);
31   cudaFree(x_d);
32   cudaMemcpy(y, y_d, sizeof(float)*N,
33              cudaMemcpyDeviceToHost);
34   cudaFree(y_d);
35 }
```

The Matlab native functions that can operate on GPUs are very few in number and since they have to be somewhat generically written they are unlikely to achieve performance on the level of hand-tuned CUDA code. On the other hand, the latter type of integration between CUDA and Matlab still requires the user to provide a CUDA implementation of the algorithm.

NVIDIA's Thrust library [6] provides a high level C++ programming interface for GPU program development. The interface resembles that of the widely known C++ standard template library and allows for the programmer to focus on implementing the core algorithm functionality, while using invocations to Thrust to perform appropriate memory transfers and thread initiation.

While having the potential to improve productivity in programming hybrid architectures, the previous approaches do not address legacy code. They are better suited for writing new applications, and to use these tools requires that existing code be rewritten, which requires significant additional effort. Other tools, such as OpenACC [8] and ExtractKernel [2], are targeted towards the porting of existing code to use the GPU.

The OpenACC standard, recently introduced by a group of computer software and hardware vendors, defines a novel method to execute unmodified C or Fortran code on a GPU. OpenACC provides special directives (or pragmas) that should be placed around parts of the code intended to execute on the GPU. These directives describe the number of threads and other configuration parameters, while the difficult task of adding code to copy CPU to GPU memories and launch GPU threads is left up to the compiler. The directives are disguised in a way that does not affect the compilation of the annotated code by compilers that do not support them. Several compilers supporting this standard are due to be released by the vendors in the first quarter of 2012.

9.3.1   ExtractKernel *Refactoring*

Our EXTRACT KERNEL refactoring is more appropriate for development based on a legacy code base, and it integrates as a plugin to the popular Eclipse integrated development environment. This refactoring is intended to serve as a rapid initial step in transitioning a legacy application to use CUDA and efficiently execute on the GPU.

EXTRACT KERNEL transforms sequential C loops into parallel CUDA code (i.e., Figs. 9.1 and 9.2). This refactoring does not abstract away CUDA code via some new programming model. On the contrary, the user of EXTRACT KERNEL will likely need to modify and maintain the generated CUDA code. Therefore, the philosophy of this refactoring is to generate code that closely follows the code style of a human CUDA developer. To accomplish this we attempt to make the parallel code within the kernel follow the contents of the refactored loop and avoid transformations that obfuscate the original code's structure.

The workflow of this refactoring is depicted in Fig. 9.3. The refactoring begins when the user concurrently selects a loop (e.g., lines 7–10 in Fig. 9.1) and chooses the "Extract to CUDA Kernel" menu option in the Eclipse environment. The selected loop is extensively analyzed to determine whether it passes a set of preconditions specific to this refactoring (box 1 in Fig. 9.3). If the preconditions are met, an initial screen is presented to the user that enables the selection of the kernel name and the tuning of GPU platform parameters (e.g., the maximum number of threads per block) (box 2a). This is followed by a refactoring preview screen that clearly and in graphical form outlines the modifications to the original code (box 3). Once the user gives his or her final approval, the refactoring takes place. All refactorings in Eclipse are easily reversible if the user is not satisfied with the end result. Also, if a candidate loop fails to pass one of the preconditions, the user is presented with an informative error message detailing the reason for rejecting the loop (e.g., the loop contains a data dependence on a specific variable) (box 2b). Once the user fixes this problem, the refactoring can be reinitiated.

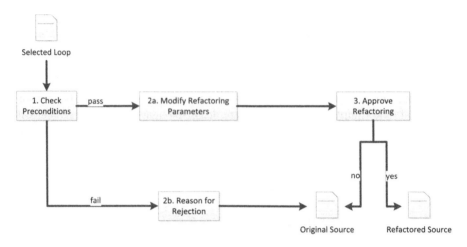

Fig. 9.3 The workflow of the EXTRACTKERNEL refactoring tool

Before the refactoring is performed, candidate loops are evaluated to determine whether they satisfy EXTRACT KERNEL's preconditions. These will only be satisfied by some of the candidate loops: ones that are parallelizable and conform to the set of rules imposed by the CUDA programming model. In order to determine whether a loop is safely parallelizable by EXTRACT KERNEL, an analysis step that determines whether the loop contains data races is performed. Intuitively, it requires that the data (i.e., array element) which is written in one iteration the loop body is not read or written from within another iteration. *Extract Kernel* uses the polyhedral model of loop iteration spaces and array accesses, which is commonly used in modern optimizing compilers, to determine the parallelizability of a loop. This approach is effective, but is necessarily conservative in that it rejects loops that may perhaps be safe, but whose analysis is inconclusive in determining their safe parallelization. Our initial preliminary tests of candidate loops, performed over large scientific codes, indicate that EXTRACT KERNEL is effective in only about half of loops. In most of the cases loops were rejected due to violating one of the preconditions (i.e., not being parallelizable or violating another CUDA requirement) and not due to a false negative induced by the loop analysis.

9.4 Conclusions

In this paper we survey several programming techniques aimed at harnessing the computational power of GPUs. These techniques range from low level extensions of C, such as CUDA or OpenCL, to high-level libraries and languages, such as Matlab and Thrust to tools like OpenACC and ExtractKernel, which are capable of adapting existing (or legacy) code to execute on GPUs.

References

1. Carver, J.C., Kendall, R.P., Squires, S.E., Post, D.E.: Software development environments for scientific and engineering software: A series of case studies. In: ICSE '07: Proceedings of the 29th International Conference on Software Engineering, pp. 550–559. IEEE Computer Society, Washington, DC, USA (2007)
2. Damevski, K., Muralimanohar, M.: A refactoring tool to extract gpu kernels. In: Proceedings of the 4th Workshop on Refactoring Tools, WRT '11, pp. 29–32. ACM, New York, NY, USA (2011)
3. Faulk, S., Loh, E., Vanter, M.L.V.D., Squires, S., Votta, L.G.: Scientific computing's productivity gridlock: How software engineering can help. Comput. Sci. Eng. **11**, 30–39 (2009)
4. Han, S., Jang, K., Park, K., Moon, S.: Packetshader: A gpu-accelerated software router. In: Proceedings of the ACM SIGCOMM Conference, pp. 195–206. ACM, New York, NY, USA (2010)
5. Kaspersky Lab utilizes NVIDIA technologies to enhance protection: URL http://www.kaspersky.com/news?id=207575979 (2009). Press release
6. Library, N.T.G.: http://code.google.com/p/thrust (2012). Accessed Dec 2012
7. Nere, A., Hashmi, A., Lipasti, M.: Profiling heterogeneous multi-gpu systems to accelerate cortically inspired learning algorithms. In: Proceedings of the IEEE International Parallel and Distributed Processing Symposium (IPDPS 2011) Anchorage, AK (2011)
8. OpenACC: http://www.openacc-standard.org (2012). Accessed Feb 2012
9. Squires, S., Van De Vanter, M., Votta, L.: Yes, there is an 'expertise gap' in hpc application development. In: Proceedings of the 3rd International Workshop on Productivity and Performance in High-End Computing (PPHEC '06). IEEE CS Press, Austin, TX (2006)

Chapter 10
Association Studies of Racial Disparities in Cancer Survivability

Lisa Walls and Weidong Mao

Abstract Cancer has always been and still is very prevalent in today's society. It would be very difficult to find a person who has not been affected by some form of the disease as either diagnosed themselves or a family member or friend diagnosed. In many families, cancer may seem to be an epidemic, constantly in some form attacking. As with many diseases, advanced research, alternative treatments, and many other factors have played major roles in increasing the survival rates of patients with cancer. In this study, we analyze different types of cancer such as breast, colon rectal, and respiratory with important attributes of patients such as age, sex, tumor size, year of treatment, and year of survival. Our results show that there is disparity between White Americans and Black Americans in cancer survivability. This can assist in predicting survivability rates and treatment of future patients.

Keywords Disparity • Association • Cancer • Survivability

10.1 Introduction

Cancer refers to any one of a large number of diseases characterized by the development of abnormal cells that divide uncontrollably and have the ability to infiltrate and destroy normal body tissue. Cancer also has the ability to spread throughout your body. Cancer is the second leading cause of death in the United States. But survival rates are improving for many types of cancer, thanks to improvements in cancer screening and cancer treatment [1].

L. Walls • W. Mao (✉)
Department of Mathematics and Computer Science, Virginia State University,
Petersburg, VA 23806, USA
e-mail: lwalls84@aol.com; wmao@vsu.edu

B. Toni et al. (eds.), *Bridging Mathematics, Statistics, Engineering and Technology*,
Springer Proceedings in Mathematics & Statistics 24, DOI 10.1007/978-1-4614-4559-3_10,
© Springer Science+Business Media New York 2012

It is difficult to find a person who has not been affected by cancer as either a victim or as a family member or friend of someone diagnosed. Often times we ponder many of the same thoughts as others. Many of these questions include:

– Is cancer hereditary or genetically linked to a family?
– How can I prevent or reduce the risk of getting cancer?
– Are there factors or circumstances in my lifestyle that will increase my chance of getting cancer?
– If I get cancer, what is my chance of survival and how can I beat the odds and increase my survival rate?

Studies have shown that in certain types of cancer there may be genetic links that when researched further have been found as a common link throughout families with similar diagnosis. Since we cannot change our genetic makeup, many times we are faced with dealing with many of the diseases and medical conditions that have attacked our family members throughout decades.

Cancer survival rates or survival statistics tell you the percentage of people who survive a certain type of cancer for a specific amount of time. Cancer statistics often use an overall five-year survival rate. For instance, the overall five-year survival rate for bladder cancer is 80%. That means that of all people diagnosed with bladder cancer, 80 of every 100 were living five years after diagnosis. Conversely, 20 out of every 100 died within five years of a bladder cancer diagnosis. Cancer survival rates are based on research that comes from information gathered on hundreds or thousands of people with a specific cancer. An overall survival rate includes people of all ages and health conditions who have been diagnosed with cancer, including those diagnosed very early and those diagnosed very late [2].

With the vast amount of data now being collected for patients diagnosed with diseases such as cancer, we would think that somewhere in the demographic or clinical data or combination thereof, there exist certain attributes that can be identified to assist in the prediction of survivability. If a set of attributes such as age, treatment type (i.e., surgery, radiation, chemotherapy, combination treatments, no surgery, etc.) can be determined that will identify the characteristics of patients with long survival, patients and doctors can use these findings to determine the plan of treatment with the best possible outcome. Association studies is the process of analyzing data from different perspectives to determine if correlations or patterns exist among a number of attributes that identify the subject. Data mining tools are able to handle large volumes of data, efficiently finding associations, patterns, and relationships that can show historical as well as future trends best on the data mining algorithm used.

Statistics shows certain cancers are more aggressive and invasive in African Americans versus White Americans. As mentioned in the Cancer Health Disparities report, some key points that attribute to mortality rates between Blacks and Whites are low socioeconomic status (SES). SES is most often based on a person's income, education level, occupation, and other factors, such as social status in the community and where he or she lives. Studies have found that SES, more than race or ethnicity, predicts the likelihood of an individual's or a group's access to education, certain occupations, health insurance, and living conditions—including conditions where exposure to environmental toxins is most common—all of which are associated with the risk of developing and surviving cancer. SES, in particular, appears to play a major role in influencing the prevalence of behavioral risk factors for cancer (e.g., tobacco smoking, physical inactivity, obesity, excessive alcohol intake, and health status), as well as in following cancer screening recommendations. Research also shows that individuals from medically underserved populations are more likely to be diagnosed with late-stage diseases that might have been treated more effectively or cured if diagnosed earlier. Financial, physical, and cultural beliefs are also barriers that prevent individuals or groups from obtaining effective health care [3].

Overall, studies show that race is a major contributor in cancer incidence as well as the increase in deaths from the disease based on race. In this paper we analyze data from breast, colon rectal, and respiratory cancer patients. With the assistance of Oracle Data Miner Software, the data are mined using the association rule to discover the race disparity in cancer survivability, specifically the commonalities of the longest survivors such as surgery/non-surgery, radiation, tumor size, race, sex between African Americans versus White Americans. This can assist in predicting survivability rates and treatment of future patients.

10.2 Methods

The dataset we use in our study comes from Surveillance Epidemiology and End Results (SEER) [4], which includes over 100 attributes that have been collected by the registry from breast cancer, colon rectal cancer and respiratory (lung) cancer. Examples of the attributes include age, year diagnosed, race, marital status, stage of cancer, tumor size, and cause of death.

In Table 10.1 we list attributes that are chosen after the initial screening according to their importance.

Table 10.2 shows the distribution of African American and White American in three different types of cancers.

The initial data mining process started with all fields loaded into the table as part of the process. The thought was that race, sex, age as well as many of the other fields would be a factor in this prediction. After reviewing the counts per each attribute, it was determined that some of the data were so disproportioned that it would obscure the results if included. For example, certain cancers are more prevalent in male versus female; using sex attribute would only cause more abnormal behavior in the results.

In order to find the disparity of cancer survivability between African Americans and White Americans, we analyze the data and try to find the association between the race with all other attributes such as age, year of survival, year of diagnosis, tumor size, surgery, and radiation treatment.

Table 10.1 Attributes and descriptions

Attributes	Descriptions
PATIENT-ID	Unique identifier for patients
REGISTRY-ID	Location submitting the data
RACE	Race of patient
SEX	Male, female
AGE-AT-DIAG	Age at the time of diagnosis
YEAR-DIAG	Year of diagnosis
CS-TUMOR-SIZE	Size of tumor
NO-SURGERY-REASON	Reason surgery was not performed
RX-RADIATION	Type of radiation performed, if any
AGE-RECODE	5 year grouping of age from 0 to 85+
RACE-RECODE	Groups race into Black, White or Other
STATE-COUNTY-RECODE	Location
SURVIVAL-TIME-RECODE	Number of months and years patient lived after diagnosis
SEER-SPEC-DEATH-CLASS	Did patient die from cancer or other cause

Table 10.2 Distributions of patients

	White	Black	Total
Breast	878,350	73,055	951,405
Respiratory	740,683	89,878	830,561
Colon/Rectal	605,229	65,242	670,471
Total	2,224,262	228,175	2,452,437

10.3 Results

Figures 10.1–10.3 show the survival percentage of Black/White diagnosed with breast cancer, colon/rectal cancer, and respiratory cancer over survival years from their diagnosis. We found that for breast cancer, in the first 5 years Blacks survive at a greater percentage than Whites, but at around 5 years and above, Whites survive longer than Blacks. For colon/rectal cancer, in the earlier years Blacks survive at a greater percentage than Whites, but Blacks surviving rates decline at a faster rate than Whites. By year 6, Whites are survival longer than Blacks. For respiratory cancer, there is no significant difference between Blacks and Whites. The majority of patients diagnosed, regardless of race, died within the first year.

Figure 10.4 shows the tumor size and percentage of patients with that size tumor who survived one year or less. We found that Black patients are usually diagnosed in later stages of their cancer which usually results in lowered survival rates. As pictured, about 27% of Blacks had tumor sizes of over 50 mm (i.e., size of a tennis ball, peach, or apple) at the time of diagnosis, while only 16% of Whites had a tumor over 50 mm. The highest percentage per tumor size for White patients is 11–20 cm at 27% (i.e., size of a penny, grape, peanut). Also, Fig. 10.4 shows that Blacks with tumor sizes of 0–20 cm had a lower survivor rate than Whites with the same tumor size. Fifteen percent of White patients survived with tumor size of 0–10 cm versus only 10% of Black patients.

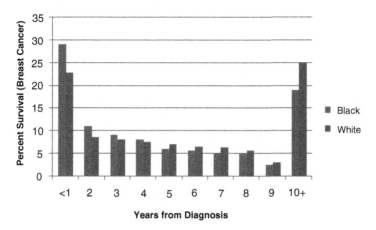

Fig. 10.1 Survival rate of breast cancer over years from diagnosis

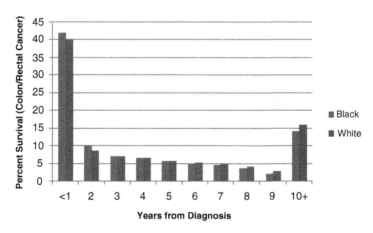

Fig. 10.2 Survival rate of colon/rectal cancer over years from diagnosis

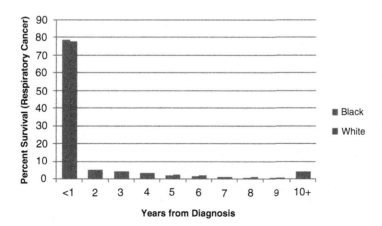

Fig. 10.3 Survival rate of respiratory cancer over years from diagnosis

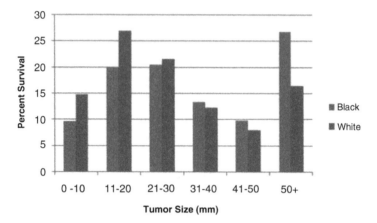

Fig. 10.4 One year or less Survival rate of breast cancer over tumor sizes

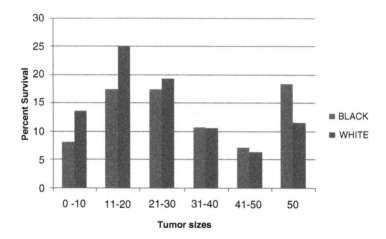

Fig. 10.5 Survival rate of breast cancer with surgery over tumor sizes

Figures 10.5 and 10.6 show the survival rate of patients for breast cancer with surgery and tumor sizes. We found that Blacks who had surgery with larger tumors (50+ mm) survived at a higher support count than Whites (Whites 11%, Blacks 18%). Blacks who had surgery with smaller tumors (0–20) had a lower survival rate than Whites with tumors of the same size (Whites 13–24%, Blacks 8–17%). NO SURGERY results for both races were about the same with less than 5% survival with all tumors except 50+ mm which resulted in a slightly higher (approximately 3%) chance of survival.

Figure 10.7 shows the one year survival rate of patients for breast cancer with treatments. We found that surgery is performed on Whites more often than Blacks. There is a approximately a 5% difference in the number of patients receiving surgery by race. (White 85%, Black 79%). Blacks had a higher survival rate over Whites

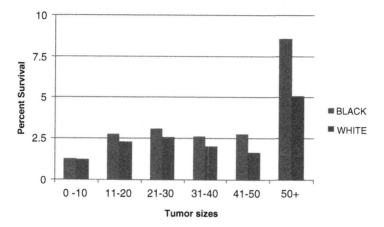

Fig. 10.6 Survival rate of breast cancer without surgery over tumor sizes

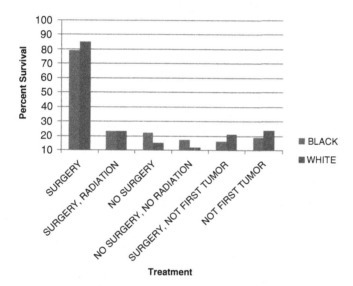

Fig. 10.7 One year survival rate of breast cancer with treatments

when No Surgery and No Radiation. This could possibly due to biological factors the make races different from others. (White 12%, Blacks 17%). Though Whites had a higher percentage of cases where the patient had multiple tumors (Not First Tumor), Blacks had a lower survival rate. (White 21%–Black 16%). We compare the five year breast cancer survivor rates with treatment in Fig. 10.8. The survival rates for Black and White counts are nearly equal for treatment of cancer. In all cases, whether surgery, no surgery, etc., the percent support remained the same among races. When compared to the one year survival we found that for five year survivors

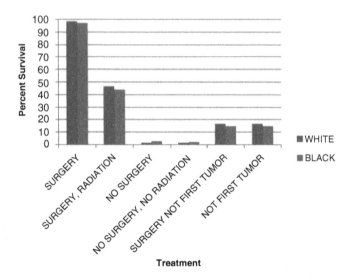

Fig. 10.8 Five year survival rate of breast cancer with treatments

the percentage survived with surgery for all categories increased from as low as 14% to as high as 40%. Five year survivors with No Surgery barely existed, whereas with one year survivors the rates were 15–22%.

The association studies for colon/rectal and respiratory cancer show very similar disparities in Black/White on survivability.

10.4 Conclusions and Future Works

In this paper we analyze the cancer patients, profile which covers about 100 attributes for three different types of cancers. Our study shows there exist disparities between African Americans and White Americans on cancer survivability. The survivability varies with factors such as year of survival, survival rates with different treatments, and tumor sizes. Other factors such as lifestyle habits (i.e., smoke, drink), living conditions, family medical history, occupation, etc. can possibly play a major role in determining the diseases we contract as well as our chances for survival.

Although these data files had hundreds of attributes, many of them were either too complex for a nonmedical analyst to decipher or spanned over so many categories that one would have to have great knowledge of the field in order to know how to appropriately group the data. If the knowledge and data were available, it would be beneficial to look at cancer stage or lymph node results to see if cancer had already spread to the other areas of the body. Knowing if the patient lived near a nuclear power plant or grew up in a big fog infested city could possibly change the outcome of the results. Future work would probably be more valuable and better represented if working alongside a medical expert with knowledge in the area of study.

Acknowledgment The project described was supported by Grant 1P20MD002731-01 from NCMHD and its contents are solely the responsibility of the authors and do not necessarily represent the official views of the NCMHD.

References

1. Mayo Foundation for Medical Education and Research: http://http://www.mayoclinic.com/health/cancer/DS01076
2. Mayo Foundation for Medical Education and Research: http://http://www.mayoclinic.com/health/cancer/CA00049
3. National Cancer Institute at the National Institutes of Health: www.cancer.gov/cancertopics
4. National Cancer Institute, Surveillance Epidemiology and End Results: www.seer.cancer.gov (2011)

Chapter 11
Perfect Hexagons, Elementary Triangles, and the Center of a Cubic Curve

Raymond R. Fletcher III

Abstract If six points in the plane are labeled with Z_6 so that for each k in Z_6 the set of lines $W_k = \{(a,b) : a+b = k\}$ concurs at a point X_k then the six points form a *perfect hexagon P*. The vertices of P and the *perspective points* $\{X_k : k \in Z_6\}$ lie on a cubic curve. If we complete *P* by including all lines which join vertices of *P* as well as all intersection points of these lines, we obtain a figure which contains many perfect hexagons. We develop a theory of cubic curves which explains this phenomenon.

Keywords Cubic curve • Hexagon • Abelian symmetric quasigroup • Sextatic points • Flex points

11.1 Introduction

A perfect n-gon P is defined in [1] as a set of n points in the plane labeled with Z_n such that for each $k \in Z_n$ the set of lines $\{(a,b) : a+b = k(\mod n)\}$ is concurrent at a point X_k. The set $\{X_k : k \in Z_n\}$ is called the *perspective set* of P. It is shown also in [1] that the combined set of vertices and perspective points of P lie on a cubic curve α which we call the *cubic envelope* of P. If α is irreducible, then a binary operation can be defined on the nonsingular points of α by letting $a * b$ equal the third point on the line (a,b) and on α. This binary operation satisfies the axioms:

1. $x * y = y * x$
2. $x * (x * y) = y$
3. $(x * y) * (u * v) = (x * u) * (y * v)$.

R.R. Fletcher III (✉)
Department of Mathematics & Computer Science, Virginia State University,
Petersburg, VA 23806, USA
e-mail: rfletcher@vsu.edu

B. Toni et al. (eds.), *Bridging Mathematics, Statistics, Engineering and Technology*,
Springer Proceedings in Mathematics & Statistics 24, DOI 10.1007/978-1-4614-4559-3_11,
© Springer Science+Business Media New York 2012

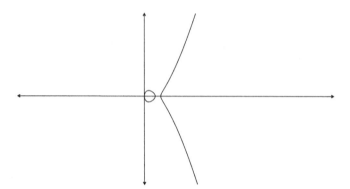

Fig. 11.1 Type IV cubic curve

In [2] an algebra satisfying axioms 1, 2 is called a *symmetric quasigroup* and an *abelian symmetric quasigroup* in case axiom 3 is satisfied also. We shall simply call an algebra satisfying (1), (2), (3) a *thirdpoint groupoid*. The element $a * a$ refers to the point of α, besides a, which lies on the tangent to α at a. A *flex* point a of α is a point at which the tangent intersects α with multiplicity 3. At such a point $a * a = a$ and thus flex points of α correspond to idempotent elements in the thirdpoint groupoid $(\alpha, *)$. In [3] it is shown that every nonsingular irreducible cubic has exactly three flex points in the real projective plane and that these are collinear.

Also in [3] it is shown that every irreducible cubic can be transformed into one of the following basic types:

   I. $y^2 = x^3$
  II. $y^2 = x^2(x+1)$
 III. $y^2 = x^2(x-1)$
  IV. $y^2 = x(x-1)(x-w), w > 1$
   V. $y^2 = x(x^2+kx+1), -2 < k < 2.$

We shall refer to any cubic which can be transformed into one of these basic types as a cubic of that type. In [4] it is shown that cubics of Type I or II cannot serve as envelopes for perfect polygons, so we will not be concerned with these. Type III cubics have one singularity at the origin. If we remove this point, we are left with a curve similar to Type V. We therefore confine our attention to the nonsingular irreducible cubics (Types IV and V). These are illustrated in Figs. 11.1 and 11.2 respectively. A Type V cubic has one connected component, and a Type IV cubic has two connected components in the real projective plane. These we shall refer to as the *oval* and the *bell*. The product of any two points on the oval is a point on the bell, and the bell is a subalgebra of $(\alpha, *)$. We note that the transformations required to take a general Type IV or Type V cubic into the indicated basic types are collineations, so any theorem involving collinearity of points or concurrence of lines, if proved for the basic type, will remain true for all curves transformable into that type.

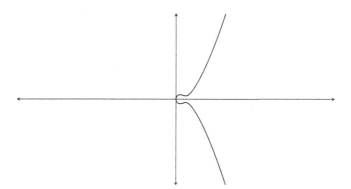

Fig. 11.2 Type V cubic curve

If P is a perfect polygon, then the *completion of P*, $(\Omega(P))$ denotes the plane figure consisting of the vertices of P, the lines joining these vertices, and the points of intersection of these lines. If $P = (0,1,2,\ldots,n-1)$ and s, t, r are fixed elements of Z_n with r, $t \neq s$, 0, we define a *cyclic derivative* $E = (E_0, E_1, \ldots, E_{n-1})$ of P by setting $E_k = (k, k+s)^\circ(k+t, k+r)$ for $k \in Z_n$. Thus a cyclic derivative of P is an n-gon whose vertices lie in $\Omega(P)$. If P is a perfect hexagon, we will show that every cyclic derivative of P is also a perfect hexagon. There are 21 cyclic derivatives of a perfect hexagon. Some of these can be easily proved perfect using the Theorems of Desargues or Pappus and others are quite difficult to prove in this way, but can be handled with analytic arguments, i.e., with appropriate transformations and coordinatizations. In this chapter we develop some properties intrinsic to irreducible cubic curves and use these to prove that the cyclic derivatives as well as many other hexagons in $\Omega(P)$ are perfect.

We shall refer to the odd subscripted perspective points of a perfect hexagon P as the *major perspective points* and the even subscripted perspective points as the *minor perspective points* of P. Three lines meet at a major perspective point and only two (not counting tangents) meet at a minor perspective point of P. Thus, to prove a hexagon is perfect it suffices to demonstrate the required concurrences at the major perspective points. We shall use the notation (a,b) to indicate the line joining points a, b; $(a,b)^\circ(c,d)$ to denote the intersection of lines (a,b) and (c,d) in the real projective plane. Also we use the notation $[a,b,c]$ as shorthand for the phrase "points a, b and c are collinear" or simply to indicate the line with points a, b, c.

11.2   The Meridians and Center of a Cubic Curve

Let α be a Type IV cubic curve with flex points e, f, g. There are two points A, B on the oval such that $A * A = B * B = f$, two points C, D on the oval such that $C * C = D * D = e$ and two points E, F on the oval such that $E * E = F * F = g$. In [5] these

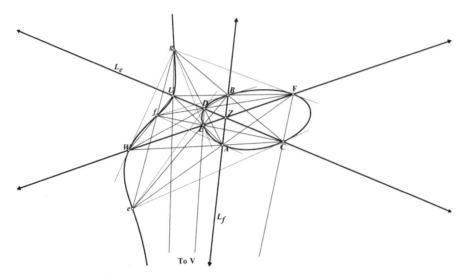

Fig. 11.3 Perfect hexagon formed by the six sextatic points

points are called the *sextatic points* of α. We denote the lines (A,B), (C,D), and (E,F) by L_f, L_e, and L_g, respectively, and call these the *meridians* of α.

Theorem 11.2.1. *The six points A, B, C, D, E, F form a perfect hexagon P with cubic envelope α. Moreover, the diagonals (A, B), (C, D), (E, F) of P are concurrent.*

Proof. Consider: $(f*C)*(f*C) = (f*f)*(C*C) = f*e = g$. It must then be that $f*C \in \{E,F\}$. Relabeling if necessary, we may suppose that $f*C = E$. Similarly $(f*D)*(f*D) = g$, and we must then have $f*D = F$. Consider $(A*e)*(A*e) = (A*A)*(e*e) = f*e = g$, and $(B*e)*(B*e) = (B*B)*(e*e) = f*e = g$. So one of $\{A*e, B*e\}$ must equal E and the other must equal F. We may suppose that $A*e = F$ and $B*e = E$ as in Fig. 11.3.

Consider $A*C = (e*F)*(f*E) = (e*f)*(E*F) = g*(E*F) = (g*g)*(E*F) = (g*E)*(g*F) = E*F$. Also $A*C = (e*F)*(f*E) = (e*E)*(f*F) = B*D$. Thus the three lines (A,C), (E,F), (B,D) concur at a point on the bell. Consider $C*F = (f*E)*(e*A) = (e*E)*(f*A) = B*A$. Also $D*E = (f*F)*(e*B) = (e*F)*(f*B) = A*B$. Thus the lines (A,B), (E,D), (C,F) concur also at a point on the bell, and similarly the lines (A,E), (C,D), (B,F) concur at a point on the bell. We have shown that the major perspective points of hexagon $P = (A,C,F,B,D,E)$ lie on α. To show that the minor perspective points of P also lie on α, consider $A*D = (e*F)*(f*F) = (e*f)*(F*F) = (e*f)*(E*E) = g*g = g = g*g = (e*E)*(f*E) = B*C$. Thus g is one of the minor perspective points and it is shown similarly that the flex points e, f are the remaining minor perspective points of P.

Now let $U = A*E = C*D = B*F$; $V = C*F = A*B = E*D$, and $W = A*C = E*F = B*D$ denote the major perspective points of P. Consider $U*V = (A*E)*(A*B) = (A*A)*(E*B) = f*e = g$ and so we have $[U,V,g]$. Let $Z = (A,B)^{\circ}(C,D)$,

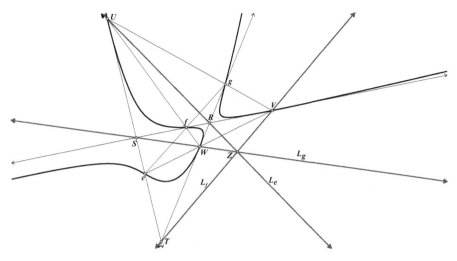

Fig. 11.4 Meridians and center of a Type V cubic

then triangles (A,E,D), (C,Z,B) are in perspective from $[U,V,g]$ and consequently the lines (A,C), (E,Z), and (D,B) are concurrent. But then Z must lie on (E,F). □

We have shown that the meridians of a Type IV cubic curve α are concurrent at a point Z. We call Z the *center* of α. A Type V cubic contains no oval so the meridians for a Type V cubic cannot be defined using the method of Theorem 11.2.1; however there is an equivalent alternative method which we now present, which can be used on both types of cubics.

Theorem 11.2.2. *Let U,V,W denote the unique nonflex points on the bell of a Type IV cubic α or a Type V cubic curve β, which square to e,f,g, respectively, and let $R = (V,f)^\circ(W,g); S = (U,e)^\circ(V,f)$, and $T = (U,e)^\circ(W,g)$. Then we define the meridians of β by $L_e = (U,R), L_f = (V,T)$, and $L_g = (W,S)$. These meridians concur at a point Z which we call the center of β. Moreover, these meridians are the same as those defined for a Type IV cubic curve in Theorem 11.2.1.*

Proof. Consider the points U,V,W defined in the proof of Theorem 11.2.1. We have $W*W = (A*C)*(D*B) = (A*D)*(C*B) = g*g = g$, and similarly $U*U = e$ and $V*V = f$. Thus the points U,V,W described in Theorems 11.2.1 and 11.2.2 are the same. Consider $(W*V)*(W*V = (W*W)*(V*V) = g*f = e$. There are only two elements on β or on the bell of α which square to e, namely $\{U.e\}$. If $W*V = U$, then we have $[W,V,U]$, but this is clearly not the case in the standard Type V cubic and thus cannot occur in any Type V cubic. So we must have $W*V = e$, and similarly $W*U = f$ and $U*V = g$ as in Figs. 11.3 and 11.4. Since $W*V = e$, we have the line $[W,V,e]$, and, referring to Fig. 11.3, triangles (g,R,f), (B,D,E) are in perspective from this line. Consequently the lines (g,B), (R,D), (f,E) are concurrent, and

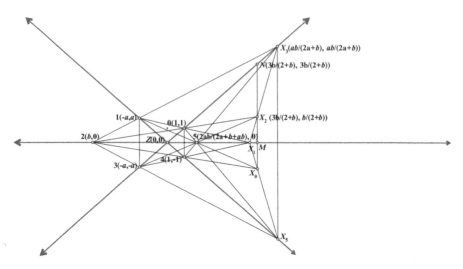

Fig. 11.5 Coordinatized perfect hexagon

since $(g,B)^\circ(f,E) = C$, we conclude that $[R,D,C]$ and thus R lies on L_e. Similarly, S lies on L_g and T on L_f. Thus the meridians as defined in Theorem 11.2.1 or Theorem 11.2.2 are the same.

Finally, for the Type V cubic illustrated in Fig. 11.4, let $Z = (R,U)^\circ(V,T)$ and consider triangles (R,S,T), (U,W,V). These are in perspective from the flex line $[e,f,g]$, with the consequence that lines (R,U), (S,W), (T,V) are concurrent. We then obtain $[Z,S,W]$ showing that the meridians of a Type V cubic curve must also be concurrent.   □

Theorem 11.2.3. *Let α_0 denote the standard Type IV cubic and β_0 the standard Type V cubic curve (see Figs. 11.1 and 11.2). In α_0 or β_0 the vertical distance from the x-axis to the flex point f is 1/3 the vertical distance from the x-axis to oblique meridian L_e.*

Proof. First consider α_0 and let $P = (0,1,2,3,4,5)$ denote the perfect hexagon on α_0 whose long diagonals are formed by the three meridians. See Fig. 11.5. Apply a horizontal translation to put the center Z of P at the origin and then a horizontal stretch/compression to put vertices 0, 4 on the vertical line $x = 1$. Finally apply a vertical stretch/compression to put these vertices at $(1,1)$ and $(1,-1)$, and let α_1 denote the resulting cubic curve. The perspective points X_0, X_2, X_4 represent the flex points e, f, g, respectively, with $X_4 = g$ the point at infinity on vertical lines. The flex line $[e,f,g]$ is vertical and meets the x-axis at M and the meridian L_e at the point N in Fig. 11.5. The transformations which carry α_0 to α_1 do not affect the ratio of MX_2 to MN, so it suffices to show that the 1:3 ratio holds in α_1. The oblique meridians L_e, L_f of α_1 have equations $y = x$ and $y = -x$, respectively, so we may set the coordinates of vertices 1, 3 at $(-a,a)$ and $(-a,-a)$, respectively. Vertex 2 must lie on the x-axis so we give it the coordinates $(b,0)$. By intersecting lines $(1,2)$, $(0,3)$ we obtain $X_3 = (ab/(2a+b), ab/(2a+b))$, and by intersecting line $(X_3,4)$ with the

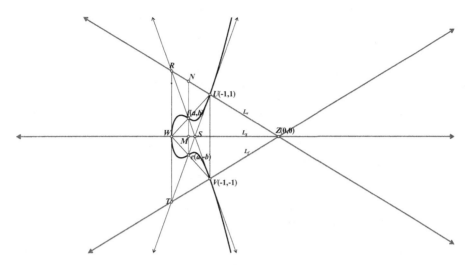

Fig. 11.6 Transformed Type V cubic curve with meridians

x-axis we obtain $(2ab/(2a+b+ab),0)$ as the coordinates of vertex 5. Finally by intersecting lines $(0,2),(3,5)$ we obtain $f = X_2 = (3b/(2+b),b/(2+b))$ and thus $MX_2 = Mf = b/(2+b)$ and $MN = 3b/(2+b) = 3(MX_2)$ as stipulated.

Now consider the standard Type V cubic curve β_0 and apply a horizontal translation to move the center Z to the origin. Let U,V,W,R,S,T be the points defined in Theorem 11.2.2, and apply a horizontal stretch/compression followed by a vertical stretch/compression to move the points U,V to $(-1,1)$ and $(-1,-1)$, respectively, as in Fig. 11.6. Let the flex points e, f have coordinates $(a,-b)$ and (a,b), respectively. By intersecting line (U,f) with the x-axis we obtain $W = ((a+b)/(1-b))$. Now intersect line (V,f) with the vertical line through W to obtain $R = ((a+b)/(1-b),2b/(1-b))$. The flex line $[e,f,g]$ is vertical and meets the x-axis at M and the meridian $L_e = (R,U)$ at N. By intersecting the flex line with (R,U) we obtain $N = (a,3b)$ and thus $MN = 3b = 3(Mf)$. $\square$

11.3   A Collineation Which Restricts to an Automorphism

If e is a flex point of an irreducible cubic curve α, then the mapping $\psi\colon \alpha \to \alpha$ defined by $\psi(x) = e*x$ is easily seen to be an automorphism of α: If x,y are points on α, then $\psi(x*y) = e*(x*y) = (e*e)*(x*y) = (e*x)*(e*y) = \psi(x)*\psi(y)$. The point $e*x$ is the preimage of x under ψ, so ψ is onto, and if $\psi(x) = \psi(y)$, then $e*x = e*y$; $e*(e*x) = e*(e*y)$, and thus $x = y$. So ψ is one–one. If x,y,z are three collinear points on α, then $x*y = z$ and $\psi(x)*\psi(y) = \psi(x*y) = \psi(z)$, and thus $[\psi(x),\psi(y),\psi(z)]$. This shows that ψ is also a collineation. In this section we show that such a mapping can be extended to a collineation of the entire plane.

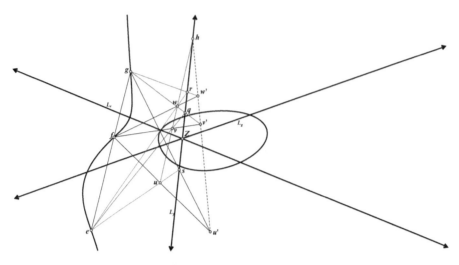

Fig. 11.7 ϕ_f is a collineation

Theorem 11.3.1. *Let α be an irreducible cubic curve with flex points e,f,g and corresponding meridians L_e, L_f, L_g. Define a mapping ϕ_f on the plane by $y = \phi_f(x) = (g,q)^\circ(f,x)$, where $q = (e,x)^\circ L_f$. Then (i)ϕ_f is an involution; (ii) ϕ_f is a collineation; (iii) ϕ_f restricts to an automorphism of α; (iv) the points fixed by ϕ_f consist of the points on L_f along with the flex point f; (v) if a,b,c,d lie on α and $(a,b)^\circ(c,d) = x$, then $\phi_f(x) = (f*a, f*b)^\circ(f*c, f*d)$; (vi) ϕ_f maps points on L_e to L_g and points on L_g to L_e.*

Proof. In Figs. 11.7 and 11.8, items (ii), (v) are illustrated on a typical Type(IV) cubic curve. However, for proving the theorem we transform to the standard Type(IV) cubic α_0 which is symmetrical w.r.t the x-axis. We then apply further transformations to place the center Z at $(0, 0)$ and the flex points e, f at $(1, -1/3)$ and $(1, 1/3)$, respectively. In accordance with Theorem 11.2.3, the meridians are then given by $L_e : y = x; L_f : y = -x$, and $L_g : y = 0$ as in Fig. 11.9. Letting $x = (s,t)$ be an arbitrary point in the plane, we find that the x-coordinate of $q = (e,x)^\circ L_f$ is given by

$$q_x = (s + 3t)/(3s + 3t - 2).$$

Since g is the point at infinity on vertical lines, this is also the x-coordinate of $y = \phi_f(x)$. Now intersecting the line (f,x) with the vertical line through q we obtain also the y-coordinate of y, and thus:

$$\phi_f(x) = \phi_f(s,t) = ((s+3t)/(3s+3t-2), (s-t)/(3s+3t-2)). \qquad (11.1)$$

In case $s+t = (2/3)$, the coordinates of $\phi_f(x)$ are not defined by (11.1) and we take $\phi_f(x)$ to be the point at infinity on lines parallel to (f,x). Using (11.1) we easily obtain $\phi_f(\phi_f(x)) = x$, and thus ϕ_f is an involution. Items (iv), (vi) are also easily proved using (11.1).

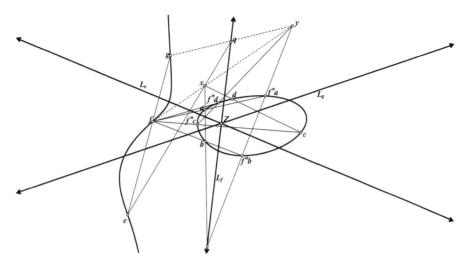

Fig. 11.8 $g = \phi_f(\mathrm{x})$

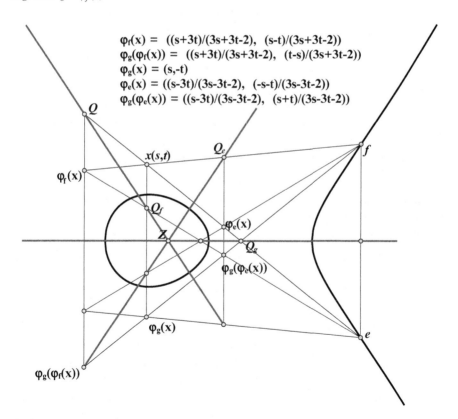

$$\varphi_f(x) = ((s+3t)/(3s+3t-2), \ (s-t)/(3s+3t-2))$$
$$\varphi_g(\varphi_f(x)) = ((s+3t)/(3s+3t-2), \ (t-s)/(3s+3t-2))$$
$$\varphi_g(x) = (s,-t)$$
$$\varphi_e(x) = ((s-3t)/(3s-3t-2), \ (-s-t)/(3s-3t-2))$$
$$\varphi_g(\varphi_e(x)) = ((s-3t)/(3s-3t-2), \ (s+t)/(3s-3t-2))$$

Fig. 11.9 The mappings φ_f, ϕ_y and φ_e

We can prove (ii) without coordinates as follows. Let u, v, w be three collinear points in the plane and let us use the notation $\phi_f(x) = x'$. Let $h = (v, w)^\circ (v', w'); k = (u, w)^\circ (u', w'); q = (e, v)^\circ L_f; r = (e, w)^\circ L_f$, and $s = (e, u) \lfloor L_f$, as in Fig. 11.7. Triangles $(g, v', w'), (e, v, w)$ are in perspective from f; thus $[q, h, r]$ and so h lie on L_f. Triangles $(g, w', u'), (e, w, u)$ are also in perspective from f, and thus $[r, k, s]$, with the consequence that k lies on L_f. So $k = L_f \lfloor (u, w) = L_f \lfloor (v, w) = h$. We then have: $[h, v', w']$ and $[k, u', w'] = [h, u', w']$. Combining, we obtain $[u', v', w']$.

To prove (iii), we first show that ϕ_f maps an arbitrary cubic curve to a cubic curve. If we let $\phi_f(s, t) = (s', t')$, then since ϕ_f is an involution, $\phi_f(s', t') = (s, t)$.

Thus $s = (s' + 3t')/M$, and $t = (s' - t')/M$, where $M = 3s' + 3t' - 2$. Now suppose (s, t) lies on the cubic curve $F(x, y) = 0$. Then $F(s, t) = 0$ implies:

$$F((s' + 3t')/M, (s' - t')/M) = 0.$$

This is a rational expression in s', t'. Multiplying by M^3 produces a cubic polynomial with the same zero set. If we set $M = M(x, y) = 3x + 3y - 2$, then (s', t') lies on the cubic curve $G(x, y) = M^3 F((x + 3y)/M, (x - y)/M) = 0$. Now consider the meridial perfect hexagon $P = (A, E, D, B, F, C)$ illustrated in Fig. 11.3. Since A, B lie on L_f and ϕ_f fixes points on L_f by (iv), we must have that $\phi_f(A) = A$ and $\phi_f(B) = B$. By (vi) we must have that $\phi_f(E) = C; \phi_f(C) = E; \phi_f(D) = F$ and $\phi_f(F) = D$. Since ϕ_f is a collineation by (ii), we must have $\phi_f(g) = \phi_f((B, C) \lfloor (A, D)) = (B, E) \lfloor (A, F) = e$. Similarly $\phi_f(f) = f$ and $\phi_f(e) = g$. Also $\phi_f(W) = \phi_f((B, D) \lfloor (A, C)) = (B, F) \lfloor (A, E) = U$, and similarly $\phi_f(U) = W$ and $\phi_f(V) = V$. Thus the 12 points on α which represent the vertices and perspective points of the meridial hexagon on α are mapped by ϕ_f to the same set of 12 points on α. Let α' denote the image of α under ϕ_f. We have shown above that α' must be a cubic curve. The 12 points lie on both α, α', and ten such points, not to speak of 12, determine a unique cubic, (see [3]), so we must have $\alpha = \alpha'$. If x is any point on α then $\phi_f(x)$ must lie on α and on the line (f, x). Thus $\phi_f(x) = f * x$. We have shown in the opening paragraph of this section that such a mapping is an automorphism of $(\alpha, *)$.

Item (v) follows immediately from (ii).   □

Let $Q_f = (g, x)^\circ L_f$; $Q_e = (f, x)^\circ L_e$, and $Q_g = (e, x)^\circ L_g$, and define mappings:

$$\phi_f(x) = (e, Q_f)^\circ (f, x),$$

$$\phi_e(x) = (g, Q_e)^\circ (e, x),$$

$$\phi_g(x) = (f, Q_g)^\circ (g, x),$$

Although ϕ_f is defined differently here than in Theorem 11.3.1, it is the same mapping. This can easily be shown using the transformed and coordinatized set up used in the proof of Theorem 11.3.1. Properties analogous to those proved for ϕ_f also hold for ϕ_e and ϕ_g, and these three mappings generate a group G of collineations

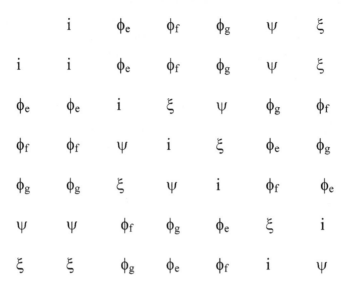

Fig. 11.10 Group of collineations

of the plane isomorphic to the symmetric group S_6. The composition table for G is given in Fig. 11.10, where $\psi = \phi_g \phi_f$ and $\xi = \phi_f \phi_g$. The coordinates given in Fig. 11.9 can be used to establish the collinearites needed to prove the table entries.

11.4 Elementary Triangles

The mapping $\psi = \phi_g \phi_f$ defined in Sect. 11.3 is a composition of two collineations, and so is itself a collineation of the plane. From the table in Fig. 11.10, it is seen that $\psi^3(x) = x$. We call $(x, \psi(x), \psi^2(x))$ the *elementary triangle* determined by the point x. We will show that the alternate vertices of any perfect hexagon with irreducible cubic envelope α form an elementary triangle. If a, b, c are three points on α such that: $a * a = b * c; b * b = a * c$, and $c * c = a * b$, then we call (a, b, c) a *perfect triangle* on α. If $P = (0, 1, 2, 3, 4, 5)$ is a perfect hexagon with cubic envelope α, then $(0, 2, 4)$ and $(1, 3, 5)$ are perfect triangles. If a is any point on α then, in [4], it is shown that there exists a unique *contiguous* perfect n-gon P on α which has a as a vertex. (If α is a Type IV cubic, "contiguous" means that all the vertices of P lie together on the bell, or they all lie together on the oval.) A perfect triangle is simply a perfect 3-gon and must be contiguous. Thus for any point a on α there exists a unique perfect triangle on α which contains a.

Theorem 11.4.1. *Every perfect triangle on an irreducible cubic curve α is an elementary triangle.*

Proof. Let a be any point on α. We will show that the elementary triangle $(a, \psi(a), \psi^2(a))$ is a perfect triangle. Let $b = \psi(a)$ and $c = \psi^2(a)$. In what follows, we use the fact that ψ being the composition $\phi_g\phi_f$ of two collineations, which restrict to automorphisms of α, is also a collineation of the plane which restricts to an automorphism of α. Consider $b*c = \psi(a)*\psi^2(a) = \psi(a*\psi(a)) = \psi(a*(g*(f*a))) = \psi(a)*(\psi(g)*(\psi(f)*\psi(a))) = \psi(a)*(f*(e*\psi(a))) = (g*(f*a))*(f*(e*\psi(a))) = (f*(f*a))*(g*(e*\psi(a))) = a*\phi_g\phi_f\psi(a) = a*a$. The requirements $c*a = b*b$ and $a*b = c*c$ follow immediately by applying ψ and ψ^2 to the equation $b*c = a*a$. Thus $(a,\psi(a),\psi^2(a))$ is the unique perfect triangle containing a. $\qquad\square$

Theorem 11.4.2. *Let P be a perfect hexagon with irreducible cubic envelope α. Then the alternate vertices of P form elementary triangles.*

Proof. Let $P = (0,1,2,3,4,5)$. Then $0*2 = 4*4 = X_2$; $0*4 = 2*2 = X_4$, and $2*4 = 0*0 = X_0$, and thus $(0,2,4)$ is a perfect triangle. Similarly $(1,3,5)$ is a perfect triangle on α. By Theorem 11.4.1, $(0,2,4)$ and $(1,3,5)$ are elementary triangles. $\qquad\square$

Since an elementary triangle is ultimately determined by an irreducible cubic curve α, we shall refer to such a triangle as α-*elementary*. Two α-*elementary* triangles are *incident* if a vertex of one lies on a side of the other. If this is the case then all three vertices of one lie one each on the three sides of the other. For, if $[x,y,\psi(x)]$, then since ψ is a collineation, we have $[\psi(x),\psi(y),\psi^2(x)]$ and $[\psi^2(x),\psi^2(y),x]$.

Theorem 11.4.3. *Let α be an irreducible cubic curve; let $\psi(x) = x'$; and let $(x,x\prime,x'')$, $(y,y\prime,y'')$ be any two nonincident α-elementary triangles. Then (x,y,x', y',x'',y'') is a perfect hexagon.*

Proof. We first transform α as in the proof of Theorem 11.3.1, so that the center Z of α lies at the origin and the meridians L_e, L_f, L_g have equations $y = x$, $y = -x$ and $y = 0$, respectively, as in Fig. 11.11. Let $x = (s,t)$ and $y = (u,v)$. Equation (11.1) then gives:

$$\phi_f(x) = ((s+3t)/(3s+3t-2),(s-t)/(3s+3t-2)),$$

$$\phi_f(y) = ((u+3v)/(3u+3v-2),(u-v)/(3u+3v-2)).$$

In this set up, ϕ_g is simply a reflection across the x-axis, and thus:

$$x' = \psi(x) = ((s+3t)/(3s+3t-2),-(s-t)/(3s+3t-2)),$$

$$y' = \psi(y) = ((u+3v)/(3u+3v-2),-(u-v)/(3u+3v-2)).$$

We then obtain:

$$x'' = \psi^2(x) = ((s-3t)/(3s-3t-2),(s+t)/(3s-3t-2)),$$

$$y'' = \psi^2(y) = ((u-3v)/(3u-3v-2),(u+v)/(3u-3v-2)).$$

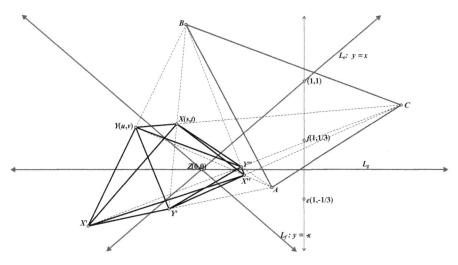

Fig. 11.11 Two elementary triangles form a perfect hexagon

Using these coordinates we find that the lines (x,y), (x',y''), (x'',y') concur at the point $C = (C_x, C_y)$ where:

$$C_x = \left(s^2u - 3t^2u - su^2 + 3t^2u^2 + 6stv - 6tuv + 3sv^2 - 3s^2v^2\right)/M,$$

$$C_y = \left(2stu + tu^2 - 3stu^2 - s^2v + 3t^2v - 2suv + 3s^2uv - 3t^2uv - 3tv^2 + 3stv^2\right)/M,$$

$$M = -s^2 - 3t^2 + 3s^2u + 3t^2u + u^2 - 3su^2 + 6stv - 6tuv + 3v^2 - 3sv^2.$$

Now let $A = \psi(C)$. Since ψ is a collineation, $[C,x,y]$ implies $[A,x',y']$; $[C,x',y'']$ implies $[A,x'',y]$, and $[C,x'',y']$ implies $[A,x,y'']$. So the lines (x',y'), (x,y''), (x'',y) concur at A. Similarly the lines (x,y'), (x',y), (x'',y'') concur at $B = \psi(A)$. Thus the hexagon (x,y,x',y',x'',y'') is perfect and its major perspective points $\{A,B,C\}$ form an elementary triangle.   □

11.5   Cyclic Derivatives and Interlaces

Two typical cyclic derivatives of a perfect hexagon are illustrated in Fig. 11.12. If the cubic envelope of a perfect n-gon P is irreducible and $n > 6$, then we do not find any perfect cyclic derivatives, but for $n = 6$ we have:

Theorem 11.5.1. *Every cyclic derivative of a perfect hexagon is perfect.*

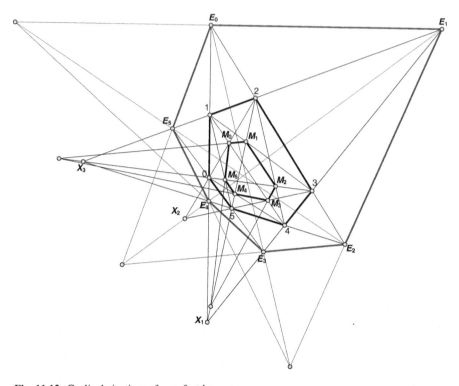

Fig. 11.12 Cyclic derivatives of a perfect hexagon

Proof. Let $P = (0,1,2,3,4,5)$ be a perfect hexagon with irreducible cubic envelope α and let $E = (E_0, E_1, E_2, E_3, E_4, E_5)$ be a cyclic derivative of P where:

$$E_k = (k, k+s)(k+t, k+r)$$

for some fixed $s, t, \ r \in Z_6$ with r, $t \neq s, 0$. Triangles $(0, 2, 4)$, $(1, 3, 5)$ are α-elementary triangles and ψ acts on the vertices of P according to the formula $\psi(x) = x + 2 \pmod 6$. Thus $\psi(E_0) = \psi((0,s)^\circ(t,r)) = (\psi(0), \psi(s))^\circ(\psi(t), \psi(r)) = (2, s+2)^\circ(t+2, r+2) = E_2$. Similarly $\psi(E_2) = E_4$ and $\psi(E_4) = E_0$. Thus (E_0, E_2, E_4) is an elementary triangle. Similarly (E_1, E_3, E_5) is an α-elementary triangle, and by Theorem 11.4.3, E is a perfect hexagon.   □

Theorem 11.5.2. *Let $P = (0,1,2,3,4,5)$ be a perfect hexagon with irreducible cubic envelope α, and let $E = (E_0, E_1, E_2, E_3, E_4, E_5)$ and $D = (D_0, D_1, D_2, D_3, D_4, D_5)$ be two cyclic derivatives of P. Then any two nonincident triangles from among $\{(E_0, E_2, E_4), (E_1, E_3, E_5), (D_0, D_2, D_4), (D_1, D_3, D_5)\}$ can be interlaced to form a perfect hexagon.*

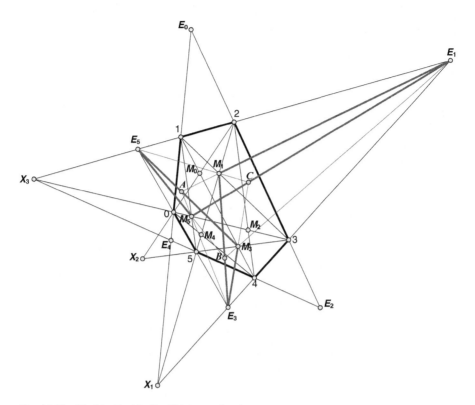

Fig. 11.13 $(E_1, M_2, E_3, M_3, E_5, M_5)$ is a perfect hexagen

Proof. As we see in the proof of Theorem 11.5.1, each of the four triangles:

$$(E_0, E_2, E_4), (E_1, E_3, E_5), (D_0, D_2, D_4), (D_1, D_3, D_5)$$

is α-elementary and so by Theorem 11.4.3 , any nonincident pair will interlace to form a perfect hexagon.   □

In Fig. 11.13 we illustrate a perfect hexagon obtained by interlacing alternate vertices from the two cyclic derivatives in Fig. 11.12. If the cubic envelope α of a perfect polygon P is reducible, it is shown in [1] that the perspective points of P lie on a line and the vertices on a conic. In [2] it is shown that every cyclic derivative of such a polygon is perfect. As we have seen, this result holds also in case P is a perfect hexagon with irreducible cubic envelope. We conjecture that if P is a perfect n-gon with $n > 6$, then P has a perfect cyclic derivative iff the cubic envelope of P is reducible.

References

1. Fletcher, R.: Perfect Polygons(1): The cubic envelope. Submitted for publication to the "Journal of Geometry" (2012)
2. Manin, Yu.: Cubic Forms. Elsevier Science Publishers B.V., New York, NY (1986)
3. Bix, R.: Conics and Cubics. Springer Science+Business Media, LLC, USA (2006)
4. Fletcher, R.: Perfect Polygons(3): Perfect Polygons with Irreducible Cubic Envelope. unpublished.
5. Walker, R.: Algebraic Curves. Springer-Verlag, New York (1978)
6. Fletcher, R.: Perfect Polygons (2): Cyclic Perfect Polygons. unpublished.

Chapter 12
Convex Quadrics and Their Characterizations by Means of Plane Sections

Valeriu Soltan

Abstract Ellipses and ellipsoids form a well-established special class of convex surfaces, primarily due to a wide range of their applications in various mathematical disciplines. The present survey deals with a natural extension of this class to that of convex quadrics. It contains a classification of convex quadrics of the Euclidean space R^n and describes, in terms of plane quadric sections, their various characteristic properties among all convex hypersurfaces of R^n, possibly unbounded.

Keywords Convex • Hypersurface • Quadratic plane • Section • Ellipsoid • Ellipse

12.1 Introduction

Various characterizations of solid ellipses and ellipsoids among convex bodes in the plane and in space became an established topic of convex geometry on the turn of the twentieth century, with Brunn's habilitation thesis [9] from 1889 and Blaschke's book [7] from 1916 reflecting significant contributions of that period. Bonnesen and Fenchel [8, § 70] gave an overview on known results in this field for dimensions two and three, published prior to 1934. Comprehensive surveys on various characteristic properties of solid ellipsoids in R^n obtained in the second half of the twentieth century are given by Gruber and Höbinger [13] and Petty [25] (see also Heil and Martini [14]). These characteristic properties usually involve one of the following topics: ellipticity or central symmetricity of plane sections of a convex body; homotheticity of parallel sections of the boundary; planarity of the sets of midpoints of families of parallel chords; planarity of shadow boundaries

V. Soltan (✉)
Department of Mathematics, George Mason University, Fairfax, VA 22030, USA.
e-mail: vsoltan@gmu.edu

B. Toni et al. (eds.), *Bridging Mathematics, Statistics, Engineering and Technology*,
Springer Proceedings in Mathematics & Statistics 24, DOI 10.1007/978-1-4614-4559-3_12,
© Springer Science+Business Media New York 2012

with respect to illuminations by parallel rays or by rays from single points; polarity of points and hyperplanes with respect to the boundary; Helmholtz–Lie type results on the mapping of the boundary of a convex body into itself.

The purpose of this survey is to describe further development of the topic, which involves quadric plane sections of convex hypersurfaces. The extension to the case of unbounded convex sets provides various characterizations of a wide class of convex quadric hypersurfaces.

The main contents of the survey are divided into the following sections:

12.2. Defining convex quadrics
12.3. Classification of convex quadrics
12.4. Quadric sections by two-dimensional planes
12.5. Quadric sections by hyperplanes

This paper is based on a talk given at the interdisciplinary Seminar on Mathematical Sciences and Applications of Virginia State University, and its contents are accessible to graduate students in mathematics.

12.2 Defining Convex Quadrics

In what follows, by *convex solids* in the Euclidean space R^n, $n \geq 2$, we mean n-dimensional closed convex sets, distinct from the whole space and, possibly, unbounded (*convex bodies* are compact convex solids). As usual, bd K and int K denote the boundary and the interior of a convex solid K.

A *convex hypersurface* in R^n is the boundary of a convex solid. This definition includes a hyperplane (i.e., a plane of dimension $n-1$) or a pair of parallel hyperplanes. A *quadric* (or a *second degree hypersurface*) in R^n is the locus of points $x = (\xi_1, \ldots, \xi_n)$ which satisfy a quadratic equation

$$F(x) \equiv \sum_{i,k=1}^{n} a_{ik}\xi_i\xi_k + 2\sum_{i=1}^{n} b_i\xi_i + c = 0, \tag{12.1}$$

where not all a_{ik} are zero.

There are different ways to define convex quadrics.

1. The most restrictive definition says that *convex quadric is a convex hypersurface in R^n which is also a quadric*. According to this definition, the convex quadrics in R^n are n-dimensional ellipsoids and paraboloids, hyperplanes, pairs of parallel hyperplanes, and all cylinders based on $(n-1)$-dimensional convex quadrics of the same type. This definition is often used in differential geometry.

2. Another way is to define *convex quadric as a convex hypersurface in R^n which is a connected component of a quadric*. This definition slightly expands the family of convex quadrics described in (1) by adding sheets of n-dimensional

elliptic hyperboloids and cylinders based on sheets of $(n-1)$-dimensional elliptic hyperboloids (see, e.g., Berger [5, Proposition 15.4.7]).

3. The definitions above exclude the boundaries of convex elliptic cones from the family of convex quadrics. The following definition from [30] corrects this omission.

Definition 12.2.1 ([30]). A convex hypersurface $S \subset R^n$ is called *convex quadric* provided there is a real quadric $Q \subset R^n$ and a connected component U of $R^n \setminus Q$ such that U is a convex set and $S = \text{bd}\, U$.

In what follows, we consider convex quadrics in R^n according to Definition 12.2.1. A classification of convex quadrics is given in Sect. 12.3.

The main topic of this survey deals with plane sections of the boundary of a convex solid. By an r-dimensional plane in R^n we mean a translate of an r-dimensional subspace, $0 \le r \le n-1$. We will say that a plane L *properly* intersects a convex solid $K \subset R^n$ (equivalently, L *properly* intersects $\text{bd}\, K$) provided L meets both sets $\text{bd}\, K$ and $\text{int}\, K$.

Lemma 12.2.2. *If an r-dimensional plane $L \subset R^n$ properly intersects a convex quadric $S \subset R^n$, then $S \cap L$ is a convex quadric in L.*

Proof. Let $Q \subset R^n$ be a real quadric and U a connected component of $R^n \setminus Q$ such that U is convex and $S = \text{bd}\, U$. First, we observe that $Q \cap L$ is a real quadric in L. Indeed, choose in R^n a new coordinate system $\eta_1, \ldots, \eta_n$, expressed by linear equations

$$\xi_i = e_{i1}\eta_1 + \cdots + e_{in}\eta_n + d_i, \quad i = 1, \ldots, n, \quad \det(e_{ij}) \ne 0, \qquad (12.2)$$

such that

$$L = \{(\eta_1, \ldots, \eta_n) : \eta_{r+1} = \cdots = \eta_n = 0\}.$$

Using (12.1) and (12.2), we describe $Q \cap L$ by a quadratic equation

$$\sum_{i,k=1}^{r} a'_{ik}\eta_i\eta_k + 2\sum_{i=1}^{r} b'_i\eta_i + c' = 0,$$

where not all scalars a'_{ik} are zero. Because $\emptyset \ne S \cap L \ne L$, the set $Q \cap L$ is a proper quadric in L. Finally, since $U \cap L$ is a convex connected component of $L \setminus (Q \cap L)$, and since $S \cap L$ is the relative boundary of $U \cap L$, the set $S \cap L$ is a convex quadric in L. □

Sections 12.4 and 12.5 contain various existing results related to the following question.

Question 12.2.3. Let $K \subset R^n$ be a convex solid and $\mathcal{F}$ a family of planes of certain dimension in R^n with the property that each plane $L \in \mathcal{F}$ properly intersects K and the set $L \cap \text{bd}\, K$ is a convex quadric in L. Describe further conditions on K and $\mathcal{F}$ to ensure that $\text{bd}\, K$ is a convex quadric.

12.3   Classification of Convex Quadrics

The next theorem from [30] plays a key role in the description of convex quadrics.

Theorem 12.3.1 ([30]). *The complement of a real quadric $Q \subset \mathbb{R}^n$, $n \geq 2$, is the disjoint union of four or fewer open sets; at least one of these components is convex if and only if the canonical form of Q is given by one of the equations*

$$a_1 \xi_1^2 + \cdots + a_k \xi_k^2 = 1, \qquad\qquad 1 \leq k \leq n,$$

$$a_1 \xi_1^2 - a_2 \xi_2^2 - \cdots - a_k \xi_k^2 = 1, \qquad\qquad 2 \leq k \leq n,$$

$$a_1 \xi_1^2 = 0,$$

$$a_1 \xi_1^2 - a_2 \xi_2^2 - \cdots - a_k \xi_k^2 = 0, \qquad\qquad 2 \leq k \leq n,$$

$$a_1 \xi_1^2 + \cdots + a_{k-1} \xi_{k-1}^2 = \xi_k, \qquad\qquad 2 \leq k \leq n,$$

where all scalars a_i involved are positive.

The proof of Theorem 12.3.1 uses the following considerations. Based on the standard classifications of quadrics $\mathbb{R}^n$ (see, e.g., Berger [5, Sect. 15.3]), we may suppose that Q has one of the following canonical forms:

$$A_k \;:\; \xi_1^2 + \cdots + \xi_k^2 = 1, \qquad\qquad 1 \leq k \leq n,$$

$$B_{k,r} \;:\; \xi_1^2 + \cdots + \xi_k^2 - \xi_{k+1}^2 - \cdots - \xi_r^2 = 1, \qquad 1 \leq k < r \leq n,$$

$$C_k \;:\; \xi_1^2 + \cdots + \xi_k^2 = 0, \qquad\qquad 1 \leq k \leq n,$$

$$D_{k,r} \;:\; \xi_1^2 + \cdots + \xi_k^2 - \xi_{k+1}^2 - \cdots - \xi_r^2 = 0, \qquad 1 \leq k < r \leq n,$$

$$E_{k,r} \;:\; \xi_1^2 + \cdots + \xi_k^2 - \xi_{k+1}^2 - \cdots - \xi_{r-1}^2 = \xi_r, \qquad 1 \leq k < r \leq n.$$

Next, we exclude the trivial cases $Q = A_1$ (when Q is a pair of parallel hyperplanes) and $Q = C_k$ (when Q is an $(n-k)$-dimensional subspace). Furthermore, the proof can be reduced to the case when Q has one of the forms $A_n, B_{k,n}, D_{k,n}, E_{k,n}$, since otherwise Q is a cylinder generated by a lower-dimensional quadric of the same type.

With this assumption, we express each of the quadrics A_n, $B_{k,n}$, $D_{k,n}$, $E_{k,n}$ as a rotation set of a respective lower-dimensional quadrics. To describe these rotations, choose any subspaces L_1, L_2, and L_3 of $\mathbb{R}^n$ such that $L_1 \subset L_2 \subset L_3$ and

$$\dim L_1 = m - 1, \;\; \dim L_2 = m, \;\; \dim L_3 = m + 1, \;\; 2 \leq m \leq n - 1.$$

Let M be the two-dimensional subspace of L_3 orthogonal to L_1. Given a point $y \in L_2$, put $M_y = y + M$ and denote by z the point of intersection of L_1 and M_y (z is the orthogonal projection of y on L_1). Let C_y be the circumference in M_y with center z

and radius $\|y - z\|$. We say that a set $X \subset L_3$ is the *rotation set* of a set $Y \subset L_2$ about L_1 within L_3 provided $X = \cup(C_y : y \in Y)$. A set $Z \subset \mathbb{R}^n$ is called *symmetric* about a subspace $N \subset \mathbb{R}^n$ if for any point $x \in Z$ and its orthogonal projection u on N, the point $2u - x$ lies in Z.

In these terms, the following lemmas hold (where $\langle e_1, \ldots, e_k \rangle$ means the span of vectors $e_1, \ldots, e_k$).

Lemma 12.3.2. *If Y is a subset of L_2 and X is the rotation set of Y about L_1 within L_3, then X is symmetric about L_2 and each component of X is the rotation set of a suitable component of Y about L_1 within L_3.*

Lemma 12.3.3. *If a set $Y \subset L_2$ is symmetric about L_1 and X is the rotation set of Y about L_1 within L_3, then X is a convex set if and only if Y is a convex set.*

Lemma 12.3.4. *Within $\mathbb{R}^n$, $n \geq 3$, we have*

(1) *A_n is the rotation set of $A_{n-1} \subset \langle e_1, \ldots, e_{n-1} \rangle$ about $\langle e_1, \ldots, e_{n-2} \rangle$,*
(2) *$B_{k,n}$ is the rotation set of $B_{k,n-1} \subset \langle e_1, \ldots, e_{n-1} \rangle$ about $\langle e_1, \ldots, e_{n-2} \rangle$, $1 \leq k \leq n - 2$.*
(3) *$D_{k,n}$ is the rotation set of $D_{k,n-1} \subset \langle e_1, \ldots, e_{n-1} \rangle$ about $\langle e_1, \ldots, e_{n-2} \rangle$, $1 \leq k \leq n - 2$.*
(4) *$B_{k,n}$ is the rotation set of $B_{k-1,n-1} \subset \langle e_2, \ldots, e_n \rangle$ about $\langle e_3, \ldots, e_n \rangle$, $2 \leq k \leq n - 1$.*
(5) *$D_{k,n}$ is the rotation set of $D_{k-1,n-1} \subset \langle e_2, \ldots, e_n \rangle$ about $\langle e_3, \ldots, e_n \rangle$, $2 \leq k \leq n - 1$.*

Finally, starting with quadric curves in $\mathbb{R}^2$, Lemmas 12.3.2–12.3.4 are used to describe recursively the quadrics in $\mathbb{R}^{m+1}$ which contain convex quadrics, based on a description of such quadrics in $\mathbb{R}^m$.

The following result from [31] gives additional details about real quadrics with convex components of the complement. We will say that a quadric $Q \subset \mathbb{R}^n$ is *proper* provided its complement $\mathbb{R}^n \setminus Q$ has two or more connected components, which happens when either Q is a hyperplane, or both sets

$$\{x \in \mathbb{R}^n : F(x) > 0\} \quad \text{and} \quad \{x \in \mathbb{R}^n : F(x) < 0\}$$

are nonempty. Furthermore, a proper quadric $Q \subset \mathbb{R}^n$ is called *locally convex* at a point $u \in Q$ if there is an open ball $U_\rho(u) \subset \mathbb{R}^n$ with center u and radius $\rho > 0$ such that $Q \cap U_\rho(u)$ is a piece of a convex hypersurface. Similarly, a proper quadric $Q \subset \mathbb{R}^n$ is called *locally supported* at $u \in Q$ provided there is an open ball $U_\rho(u) \subset \mathbb{R}^n$ and a hyperplane $H \subset \mathbb{R}^n$ through u such that $Q \cap U_\rho(u)$ lies in a closed halfspace of $\mathbb{R}^n$ determined by H.

Theorem 12.3.5 ([31]). *For a proper quadric $Q \subset \mathbb{R}^n$, $n \geq 2$, the following conditions are equivalent:*

(1) *at least one of connected components of $\mathbb{R}^n \setminus Q$ is a convex set.*
(2) *Q is locally convex at a certain point $u \in Q$.*
(3) *Q is locally supported at a certain point $u \in Q$.*

There is a connection of conditions (2) and (3) in Theorem 12.3.5 with respective properties of convex hypersurfaces. Indeed, if S is the boundary of an open connected set $X \subset \mathbf{R}^n$, then S is a convex hypersurface provided X is locally supported at every point $u \in S$ (see Carathéodory [12]). Similarly, S is a convex hypersurface if X is locally convex at every point $u \in S$ (see Nakajima [22] and Tietze [32]). On the other hand, Theorem 12.3.5 deals with local convexity and local support of Q at a *single* point.

Theorem 12.3.1 immediately implies the following classification of convex quadrics.

Corollary 12.3.6 ([30]). *A convex hypersurface $S \subset \mathbf{R}^n$, $n \geq 2$, is a convex quadric if and only if S can be described in suitable Cartesian coordinates $\xi_1, \ldots, \xi_n$ by one of the conditions:*

$$a_1 \xi_1^2 + \cdots + a_k \xi_k^2 = 1, \qquad\qquad 1 \leq k \leq n,$$

$$a_1 \xi_1^2 - a_2 \xi_2^2 - \cdots - a_k \xi_k^2 = 1, \ \xi_1 \geq 0, \qquad 2 \leq k \leq n,$$

$$a_1 \xi_1^2 = 0,$$

$$a_1 \xi_1^2 - a_2 \xi_2^2 - \cdots - a_k \xi_k^2 = 0, \ \xi_1 \geq 0, \qquad 2 \leq k \leq n,$$

$$a_1 \xi_1^2 + \cdots + a_{k-1} \xi_{k-1}^2 = \xi_k, \qquad\qquad 2 \leq k \leq n,$$

where all scalars a_i involved are positive.

In particular, convex quadrics in $\mathbf{R}^n$ which contain no lines can be expressed in suitable coordinates by one of the equations

$$a_1 \xi_1^2 + \cdots + a_n \xi_n^2 = 1, \qquad\qquad \text{(ellipsoid)}$$

$$a_1 \xi_1^2 - a_2 \xi_2^2 - \cdots - a_n \xi_n^2 = 1, \ \xi_1 \geq 0, \qquad \text{(sheet of elliptic hyperboloid}$$
$$\text{of two sheets)}$$

$$a_1 \xi_1^2 - a_2 \xi_2^2 - \cdots - a_n \xi_n^2 = 0, \ \xi_1 \geq 0, \qquad \text{(sheet of elliptic cone)}$$

$$a_1 \xi_1^2 + \cdots + a_{n-1} \xi_{n-1}^2 = \xi_n, \qquad\qquad \text{(elliptic paraboloid)}$$

where all scalars $a_1, \ldots, a_n$ are positive.

A recurrent description of convex quadratics in $\mathbf{R}^n$ can be given as follows.

1. Convex quadratic curves in $\mathbf{R}^2$ are ellipses, branches of hyperbolas, parabolas, convex cones, lines, and pairs of parallel lines.
2. Convex quadratic in $\mathbf{R}^n$, $n \geq 3$, are ellipsoids, sheets of elliptic hyperboloids of two sheets, sheets of elliptic cones, elliptic paraboloids, and cylinders based on convex quadrics in $\mathbf{R}^{n-1}$.

12.4 Quadric Sections by Two-Dimensional Planes

The following result was proved by Kubota [19] for $n = 3$ (see also Auerbach, Mazur, and Ulam [4]) and by Busemann [11, pp. 91–92] for all $n \geq 3$.

Theorem 12.4.1 ([11,19]). *If a convex body $K \subset R^n$, $n \geq 3$, has the property that every two-dimensional plane through a fixed point $p \in \operatorname{int} K$ intersects the boundary of K in an ellipse, then* $\operatorname{bd} K$ *is an ellipsoid.*

The methods of proofs in [11] and [19] are essentially different. We sketch here both methods, since their variations are widely used in the proofs of the results listed below.

Kubota chooses ellipses E_1, E_2, E_3, which are the sections of $\operatorname{bd} K$ by three distinct planes through p containing no common line, and considers a quadric Q containing $E_1 \cup E_2 \cup E_3$. Any plane L through p that meets $E_1 \cup E_2 \cup E_3$ at a set S of precisely six points determines two quadrics: $L \cap \operatorname{bd} K$ and $L \cap Q$, both containing S. Since a planar quadric curve is uniquely determined by any five points with no four on a line (see, e.g., [23]), we have $L \cap \operatorname{bd} K = L \cap Q$. Varying L about p, one easily obtains that $\operatorname{bd} K = Q$.

Busemann uses synthetic methods and geometric transformations of the space to show that $\operatorname{bd} K$ itself is an ellipsoid (although his proof employs projective transformations of R^3, we describe below its affine version). Choose a chord $[x, z]$ of the convex body $K \subset R^3$ which contains p and has maximum possible length. There are parallel planes H_x and H_z through x and z, respectively, both supporting K (see, e.g., [27]). Applying a suitable affine transformation, one may assume that $[x, z]$ is perpendicular to both H_x and H_z. Then every section of $\operatorname{bd} K$ by a plane L through $[x, z]$ is an ellipse symmetric about $[x, z]$. Let L_0 be the plane through p which is parallel to H_x. It is easy to see that $E_0 = L_0 \cap \operatorname{bd} K$ is an ellipse symmetric about p. Applying one more affine transformation that keeps $[x, z]$ and L_0 and transforms E_0 into a circle, we obtain that $\operatorname{bd} K$ is an ellipsoid of rotation about $[x, z]$.

The following statement is proved in [28] by using an extension of Busemann's method to the case of unbounded convex sets.

Theorem 12.4.2 ([28]). *If $K \subset R^n$, $n \geq 3$, is a convex solid and $p \in \operatorname{int} K$, then the boundary of K is a convex quadric if and only if all sections of $\operatorname{bd} K$ by two-dimensional planes through p are convex quadric curves.*

The next result from [31] sharpens Theorem 12.4.2 by putting restrictions on the directions of two-dimensional planes through p (see also Montejano and Morales [20] for the case of a convex body symmetric about p). We recall that the *recession cone* of a convex solid $K \subset R^n$ is defined by

$$\operatorname{rec} K = \{y \in R^n : x + \lambda y \in K \text{ whenever } x \in K \text{ and } \lambda \geq 0\}.$$

The set $\operatorname{rec} K$ is a closed convex cone with apex o, the origin of $\mathbf{R}^n$; furthermore, $\operatorname{rec} K$ is distinct from $\{o\}$ if and only if K is unbounded. The next result is obtained by using Kubota's method.

Theorem 12.4.3 ([31]). *Let $K \subset \mathbf{R}^n$, $n \geq 3$, be a convex solid and $p \in \operatorname{int} K$. Suppose l is a line through p such that the one-dimensional subspace $l - p$ does not lie in $\operatorname{rec} K \cup -\operatorname{rec} K$. For any scalar $\varepsilon > 0$, the following conditions are equivalent:*

(1) *$\operatorname{bd} K$ is a convex quadric.*
(2) *For every two-dimensional plane L through p which forms with l an angle of size ε or less, the section $L \cap \operatorname{bd} K$ is a convex quadric curve.*

Kubota's method also allows further refinement of Theorem 12.4.2. We observe that if $K \subset \mathbf{R}^n$ is a closed n-dimensional convex cone with apex p and L is a two-dimensional plane through p properly intersecting K, then $L \cap \operatorname{bd} K$ is a convex cone, which is a convex quadric.

Theorem 12.4.4. *Let $K \subset \mathbf{R}^n$, $n \geq 3$, be a convex solid and p a point in $\operatorname{bd} K$ such that all proper sections of $\operatorname{bd} K$ by two-dimensional planes through p are convex quadric curves. Then either K is a convex cone with apex p or $\operatorname{bd} K$ is a convex quadric.*

Proof. We proceed by induction on $n (\geq 3)$. Let $n = 3$. If K contains a line l and L is a plane complementary to l, then $\operatorname{bd} K$ is a cylindric surface based on the convex quadric curve $L \cap \operatorname{bd} K$. Let K contain no lines. Translating K on the vector $-p$, we may suppose that $p = o \in \operatorname{bd} K$. Assume that K is not a cone with apex o. Then there is a line l through o meeting $\operatorname{int} K$ such that $l \cap K$ is a line segment, $[o, z]$. Choose a pair of distinct two-dimensional subspaces L_1 and L_2 both containing l and such that the sets $L_1 \cap K$ and $L_2 \cap K$ are bounded. By the assumption, $E_1 = L_1 \cap \operatorname{bd} K$ and $E_2 = L_2 \cap \operatorname{bd} K$ are convex quadric curves, whence they are ellipses. Let c be the midpoint of $[o, z]$, and c_1 and c_2 the centers of E_1 and E_2, respectively. Applying a suitable affine transformation, we may assume that both E_1 and E_2 are circles and the planes L_1 and L_2 are orthogonal. Clearly, the image of K under this transformation, also denoted by K, satisfies theorem's hypothesis. Let 2δ be the length of $[o, z]$.

Choose in $\mathbf{R}^3$ a coordinate system (ξ_1, ξ_2, ξ_3) such that l is the ξ_3-axis, all points c, c_1, c_2 belong to the coordinate plane $\xi_3 = \sigma_3$, where $\sigma_3 \geq 0$ is a suitable scalar. So, we may put

$$c = (0, 0, \sigma_3), \quad c_1 = (\sigma_1, 0, \sigma_3), \quad c_2 = (0, \sigma_2, \sigma_3), \quad \sigma_1, \sigma_2, \sigma_3 \geq 0.$$

Then E_1 and E_2 are described by

$$E_1 = \{(\xi_1, 0, \xi_3) : (\xi_1 - \sigma_1)^2 + (\xi_3 - \sigma_3)^2 = \sigma_1^2 + \delta^2\},$$
$$E_2 = \{(0, \xi_2, \xi_3) : (\xi_2 - \sigma_2)^2 + (\xi_3 - \sigma_3)^2 = \sigma_2^2 + \delta^2\}.$$

Clearly, L_1 and L_2 are given by the equations $\xi_2 = 0$ and $\xi_1 = 0$, respectively.

Choose a point $v \in \operatorname{bd} K \setminus (L_1 \cup L_2)$ so close to o that a certain two-dimensional plane L through the line $\langle o, v \rangle$ meets K along a bounded set and intersects each of the ellipses E_1, E_2 at precisely three points (including o). Since K is not a cone with apex o, the point v can be chosen such that $[o, v]$ meets $\operatorname{int} K$. As above, $L \cap \operatorname{bd} K$ is an ellipse.

We state the existence of a quadric Q that contains $\{v\} \cup E_1 \cup E_2$. For this, consider the family of quadrics $Q(\mu)$ given by

$$\xi_1^2 + \xi_2^2 + \xi_3^2 + \mu \xi_1 \xi_2 - 2\sigma_1 \xi_1 - 2\sigma_2 \xi_2 - 2\sigma_3 \xi_3 + \sigma_3^2 - \delta^2 = 0,$$

where μ is a scalar parameter. Obviously, $E_i = L_i \cap Q(\mu)$, $i = 1, 2$, for all $\mu \in \mathbf{R}$. If $v = (v_1, v_2, v_3)$, then $v \notin L_1 \cup L_2$ if and only if $v_1 v_2 \neq 0$. Hence $v \in Q = Q(\mu_0)$, where

$$\mu_0 = \frac{\delta^2 - \sigma_3^2 + 2\sigma_1 v_1 + 2\sigma_2 v_2 + 2\sigma_3 v_3 - v_1^2 - v_2^2 - v_3^2}{v_1 v_2}.$$

Next, we state that $L \cap \operatorname{bd} K \subset Q$. Indeed, denote by H a plane supporting K at o. Since H supports $E_1 \cup E_2$, the plane H is uniquely defined and is tangent to Q. It is known that a planar quadric curve is uniquely determined by any four points (not all on a line) and a tangent line at one of them (see, e.g., [24]). Since both planar quadrics $L \cap \operatorname{bd} K$ and $L \cap Q$ contain the four-point set $\{v\} \cup (L \cap (E_1 \cup E_2))$ and are tangent to the line $H \cap L$ at o, they coincide by the argument above. Hence $L \cap \operatorname{bd} K = L \cap Q \subset Q$.

Slightly rotating L about the line $\langle o, v \rangle$, we obtain a family of ellipses $L \cap \operatorname{bd} K$ which cover an open subset V of $\operatorname{bd} K$. As above, $V \subset Q$. Let $w \in V$ such that $[o, w]$ meets $\operatorname{int} K$. To show the inclusion $\operatorname{bd} K \subset Q$, choose any point $x \in \operatorname{bd} K \setminus \{w\}$ and denote by N the two-dimensional subspace containing $\{w, x\}$. Since the quadric curves $N \cap \operatorname{bd} K$ and $N \cap Q$ coincide along the nonlinear arc $N \cap W$ of a quadric curve, they must coincide: $N \cap \operatorname{bd} K = N \cap Q$. Summing up, $\operatorname{bd} K \subset Q$. Because Q is locally convex at any point $x \in \operatorname{bd} K$, Theorem 12.3.5 implies that $\operatorname{bd} K$ is a convex quadric.

Let $n \geq 4$. As above, we assume that $o \in \operatorname{bd} K$. Choose a point $q \in \operatorname{int} K$ and a two-dimensional plane M through q. Consider a three-dimensional subspace S containing $\{o\} \cup L$ and the three-dimensional closed convex set $P = K \cap S$. If L is a two-dimensional subspace of S properly intersecting P, then from $L \cap \operatorname{rbd} P = L \cap \operatorname{bd} K$ and the inductive hypothesis implies that L meets the relative boundary $\operatorname{rbd} P$ of P along a convex quadric curve. By the proved above (the case $n = 3$), $\operatorname{rbd} P$ is a convex quadric in S. Hence $L \cap \operatorname{bd} K = L \cap \operatorname{rbd} P$ is a convex quadric curve, and Theorem 12.4.2 shows that $\operatorname{bd} K$ is a convex quadric. $\qquad \square$

Burton [10] and Höbinger [15, Theorems 3 and 6] independently obtained the same refinement of Theorem 12.4.1 by showing that the point p can be chosen anywhere in $\mathbf{R}^n$. In this regard we formulate the following problem, which complements Theorems 12.4.2 and 12.4.4.

Problem 12.4.5 ([30]). Is it true that the boundary of a convex solid $K \subset \mathbb{R}^n$, $n \geq 3$, is a convex quadric if and only if there is a point $p \in \mathbb{R}^n \setminus K$ such that all proper sections of $\mathrm{bd}\, K$ by two-dimensional planes through p are convex quadric curves?

Corollary 12.4.11 implies that the boundary of a convex solid $K \subset \mathbb{R}^n$, $n \geq 3$, is a convex quadric if and only if there is a point $p \in \mathbb{R}^n \setminus K$ such that all proper sections of $\mathrm{bd}\, K$ by 3-*dimensional* planes through p are convex quadrics. One more problem concerns a local version of Theorem 12.4.2.

Problem 12.4.6. Let K be a convex solid in $\mathbb{R}^n$, $n \geq 3$, and Ω an open subset of $\mathrm{bd}\, K$. Is it true that the following two conditions are equivalent?

(1) Ω lies within a convex quadric.
(2) There is a point $p \in \mathrm{int}\, K$ such that every section of Ω by a two-dimensional plane through p lies within a convex quadric curve.

Another way to extend Theorem 12.4.1 to the case of convex solids is to consider their *bounded* planar sections. We will say that a convex solid K is *line-free* if it contains no lines.

Theorem 12.4.7 ([31]). *For a line-free convex solid $K \subset \mathbb{R}^n$ and a point $p \in \mathbb{R}^n$, $n \geq 3$, the following conditions are equivalent:*

(1) *All proper bounded sections of $\mathrm{bd}\, K$ by two-dimensional planes through p are ellipses.*
(2) *The set $\mathrm{bd}\, K \setminus \big((p + \mathrm{rec}\, K) \cup (p - \mathrm{rec}\, K)\big)$ lies in a convex quadric.*

We observe that condition (1) of Theorem 12.4.7 implicitly covers the trivial case when no proper section of $\mathrm{bd}\, K$ by a two-dimensional plane through p is bounded. For the line-free convex solid K, this happens if and only if $K \subset p + \mathrm{rec}\, K$, or, equivalently, when the set $\mathrm{bd}\, K \setminus \big((p + \mathrm{rec}\, K) \cup (p - \mathrm{rec}\, K)\big)$ is empty, thus ensuring the equivalence of conditions (1) and (2) of the theorem.

The following example shows that in Theorem 12.4.7 the boundary of K can be different from a convex quadric.

Example 12.4.8. Let K be a convex solid in $\mathbb{R}^3$, is given by

$$K = \big\{(\xi_1, \xi_2, \xi_3) \mid \xi_3 \geq \max\big\{1, (\xi_1^2 + \xi_2^2)^{1/2}\big\}\big\}$$

(so that the boundary of K is a truncated sheet of a convex circular cone). Let $p = (0, 0, 2)$. Then $p \in \mathrm{int}\, K$, and a plane L through p intersects K along a bounded set if and only if L misses the open circle

$$C = \{(\xi_1, \xi_2, \xi_3) \mid \xi_1^2 + \xi_2^2 < 1,\ \xi_3 = 1\}.$$

Furthermore, all bounded sections of $\mathrm{bd}\, K$ by two-dimensional planes through p are ellipses. Clearly, $\mathrm{bd}\, K \setminus \big((p + \mathrm{rec}\, K) \cup (p - \mathrm{rec}\, K)\big)$ is the part of $\mathrm{bd}\, K$ disjoint from the plane $\xi_3 = 1$.

Two-dimensional planar sections provide characterizations of various classes of convex solids. A well-known result of Klee [16] states that a convex solid $K \subset R^n$ is a convex polyhedron (i.e., the intersection of finitely many closed halfspaces) if and only if there is a point $p \in \operatorname{int} K$ such that every two-dimensional plane through p meets K along a polygonal set. The following result from [26, 28] combines Klee's statement and Theorem 12.4.2 (see also [28] for a more general statement which involves so-called boundedly polyhedral convex solids).

Theorem 12.4.9 ([26, 28]). *The boundary of a convex solid $K \subset R^n$, $n \geq 3$, is a convex quadric or a convex polyhedral hypersurface if and only if there is a point $p \in \operatorname{int} K$ such that every section of $\operatorname{bd} K$ by a two-dimensional plane through p is a convex quadric curve or a convex polygonal line.*

The proof of Theorem 12.4.9 is organized by induction on $n \, (\geq 3)$. For the basic case, $n = 3$, we proceed by the method of contradiction. This gives the existence of two planes through p, say L_1 and L_2, such that $E_1 = L_1 \cap \operatorname{bd} K$ is a convex quadric distinct from a cone, and $E_2 = L_2 \cap \operatorname{bd} K$ is a convex polygonal line. Using the polygonality of E_2, one can find an integer m and a sequence of planes $N_1, N_2, \ldots$ through p converging to L_1 such that each section $P_i = N_i \cap \operatorname{bd} K$, $i = 1, 2, \ldots$, is a convex polygonal line with at most m sides. Since no line-free convex quadric distinct from a convex cone can be the limit of a sequence of such polygonal lines, we obtain a contradiction with the assumption $P_i \to E_1$.

The following example shows that, unlike Theorems 12.4.4 and 12.4.7 above, the point p in Theorem 12.4.9 cannot be placed in $R^n \setminus \operatorname{int} K$.

Example 12.4.10. Let $K \subset R^3$ be a truncated bounded cone, given by

$$K = \{(\xi_1, \xi_2, \xi_3) : \xi_1 \geq (\xi_1^2 + \xi_2^2)^{1/2}, \, 1 \leq \xi_1 \leq 2\},$$

and let $p = (0, 0, 0)$. Then all proper sections of $\operatorname{bd} K$ by planes through p are trapezoids, while K is neither a solid ellipsoid nor a convex polytope.

In fact, Theorem 12.4.1 is formulated by Busemann in a slightly more general form. Namely, as pointed in [11], the boundary of a convex body is an ellipsoid if and only if there is an integer r, $2 \leq r \leq n - 1$, such that every r-dimensional plane through a fixed point $p \in \operatorname{int} K$ intersects $\operatorname{bd} K$ in an r-dimensional ellipsoid (ellipse when $r = 2$). This statement was further refined by Höbinger [15], who showed that p can be selected anywhere in R^n. We observe here that the case of any r between 2 and $n - 1$ is easily reducible to that of $r = 2$. Furthermore, the case of any r between 2 and $n - 1$ can be refined as follows.

Corollary 12.4.11. *The boundary of a convex solid $K \subset R^n$, $n \geq 3$, is a convex quadric if and only if there is a plane $L \subset R^n$ of certain dimension s, $0 \leq s \leq n - 4$, and an integer r, with $s + 3 \leq r \leq n - 1$, such that all proper sections of $\operatorname{bd} K$ by r-dimensional planes through L are r-dimensional convex quadrics. If L meets K or if K is a convex body, then one can assume that $s \leq n - 3$ and $s + 2 \leq r$.*

Proof. Clearly, we have to verify the "only if" part. Let $L \subset \mathbb{R}^n$ be an s-dimensional plane and p a point in int K. Choose a two-dimensional plane through p that properly intersects K. Since $s + 3 \leq r$, there is an r-dimensional plane $M \subset \mathbb{R}^n$ containing $L \cup N$. Obviously, M properly intersects K. By the assumption, $M \cap \mathrm{bd}\, K$ is an r-dimensional convex quadric. Hence $L \cap \mathrm{bd}\, K = L \cap (M \cap \mathrm{bd}\, K)$ is a convex quadric curve. Now Theorem 12.4.2 above implies that $\mathrm{bd}\, K$ is a convex quadric.

If L meets K, then let p be a point in $L \cap K$, and if K is a convex body, then let p be any point in L. If N is a two-dimensional plane through p that properly intersects K, then $L \cup N$ lies in a $s + 2$ dimensional plane. Hence, with $s + 2 \leq r$, we can find an r-dimensional plane $M \subset \mathbb{R}^n$ containing $L \cup N$. As above, $L \cap \mathrm{bd}\, K = L \cap (M \cap \mathrm{bd}\, K)$ is a convex quadric curve. Finally, Theorem 12.4.4 (if L meets K) or Theorem 4 from [10] (if K is a convex body) shows that $\mathrm{bd}\, K$ is a convex quadric. $\qquad\square$

The following example shows that the inequality $s + 2 \leq r$ in Corollary 12.4.11 above cannot be replaced by $s + 1 \leq r$ if L meets K or if K is a convex body (see Nakagawa [21] for a similar example involving ellipses).

Example 12.4.12. Let Γ be a strictly convex quadric curve in the $\xi_1 \xi_3$-plane of $\mathbb{R}^3$ symmetric about the ξ_3-axis. Denote by Γ' the curve in the $\xi_2 \xi_3$-plane obtained from Γ by rotation about the ξ_3-axis. Let K be the convex solid in $\mathbb{R}^3$ which is the convex hull of $\Gamma \cup \Gamma'$. If L is the ξ_3-axis (so, $s = 1$), then each section of $\mathrm{bd}\, K$ by a two-dimensional plane through L is a convex quadric curve, while $\mathrm{bd}\, K$ is not a convex quadric surface.

Petty [25] observed (based on statements of Busemann [11] and Burton [10]) that the boundary of a convex body $K \subset \mathbb{R}^n$ is an ellipsoid provided for a given point p in the projective extension $\mathbb{P}^n$ of $\mathbb{R}^n$ and an integer r, $2 \leq r \leq n - 1$, all proper sections of $\mathrm{bd}\, K$ by r-dimensional planes through p are r-dimensional ellipsoids. For the case when $p \in \mathbb{P}^n \setminus \mathbb{R}^n$, Petty's observation can be reformulated as follows: the boundary of K is an ellipsoid provided every r-dimensional plane $L \subset \mathbb{R}^n$ which is parallel to a given line $l \subset \mathbb{R}^n$ and properly intersects K, the section $L \cap \mathrm{bd}\, K$ is an r-dimensional ellipsoid.

As above, we observe that the case of any r between 2 and $n - 1$ is easily reducible to that of $r = 2$. The following result (whose proof uses a modification of Kubota's method) extends Petty's statement (for $r = 2$) to the case of convex quadrics and puts additional restrictions on the planes which meet K. Given a line $l \subset \mathbb{R}^n$ and a scalar $\delta > 0$, denote by $P_\delta(l)$ the family of two-dimensional planes which are parallel to l and whose distance from l is less than δ.

Theorem 12.4.13 ([31]). *Let $K \subset \mathbb{R}^n$, $n \geq 3$, be a convex solid, l a line which meets K along a bounded set, and δ a positive scalar. The following conditions are equivalent:*

(1) *$\mathrm{bd}\, K$ is a convex quadric.*
(2) *For any plane $L \in P_\delta(l)$ properly intersecting K, the section $L \cap \mathrm{bd}\, K$ is a convex quadric curve.*

The following example shows that in Theorem 12.4.13, the condition on the line l to meet K along a bounded set is essential. Let Q be the unit square in the coordinate plane $\xi_1 = 0$ of R^3, and l be the ξ_1-axis of R^3. If K is the Cartesian product of Q and l, then $\operatorname{bd} K$ is not a convex quadric, while any proper section of $\operatorname{bd} K$ by a two-dimensional plane parallel to l consists of a pair of parallel lines, which is a degenerate convex quadric curve.

12.5 Quadric Sections by Hyperplanes

Kubota [17, 18] showed that, given a pair of bounded convex surfaces in R^3, one being enclosed by the other, if all planar sections of the biggest surface by planes tangent to the enclosed surface are ellipses, then the biggest surface is an ellipsoid. Bianchi and Gruber [6] gave a far-reaching generalization of this statement, extended in [30] to the case of convex quadrics.

Theorem 12.5.1 ([30]). *Let K be a convex solid in R^n, $n \geq 3$, and $\delta(u)$ a continuous real-valued function on the unit sphere S^{n-1} of R^n such that for each vector $u \in S^{n-1}$ the hyperplane $\{x \in R^n : x \cdot u = \delta(u)\}$ either lies in K or intersects $\operatorname{bd} K$ along an $(n-1)$-dimensional convex quadric. Then $\operatorname{bd} K$ is a convex quadric.*

Alonso and Martín [1–3] obtained a serious of characterizations of ellipsoids in R^n by means of hyperplane sections. We need some definitions to formulate their results.

Let $L \subset R^n$ be a plane of dimension r, $0 \leq r \leq n-2$. We say that a convex body $K \subset R^n$ is *elliptic through* L if for every hyperplane H containing L and properly intersecting K, the set $H \cap \operatorname{bd} K$ is an $(n-1)$-dimensional ellipsoid. For any planes P_1 and P_2 in R^n, denote by $\operatorname{aff}(L_1 \cup L_2)$ the smallest plane containing $L_1 \cup L_2$.

Theorem 12.5.2 ([2,3]). *Let K be a convex body in R^n, $n \geq 3$, and L_1 and L_2 be two planes of dimensions r_1 and r_2, respectively, $1 \leq r_1, r_2 \leq n-2$, such that $L_1 \not\subset L_2$ and $L_2 \not\subset L_1$. Assume that K is elliptic through each of L_1 and L_2 and that one of the following properties holds:*

(a) At least one of the planes L_1 and L_2 meets $\operatorname{int} K$.
(b) Both L_1 and L_2 support K, $L_1 \cap K = L_2 \cap K$, and $\dim(\operatorname{aff}(L_1 \cup L_2)) < n$.
(c) Both L_1 and L_2 support K, $L_1 \cap L_2 = \emptyset$, and $\dim(\operatorname{aff}(L_1 \cup L_2)) = n$.

Then $\operatorname{bd} K$ is an ellipsoid.

Due to Corollary 12.4.11, we may put $r_1 = r_2 = n-2$ in Theorem 12.5.2. Papers [1–3] contain a variety of sophisticated examples demonstrating the necessity of conditions (a)–(c) in Theorem 12.5.2. Following [1], we say that a convex body $K \subset R^n$ is *elliptic* with respect to an $(n-1)$-dimensional subspace H of R^n provided any proper section of $\operatorname{bd} K$ by a translate of H is an $(n-1)$-dimensional ellipsoid.

Theorem 12.5.3 ([1]). *If a centrally symmetric convex body $K \subset \mathbb{R}^n$ is elliptic with respect to three pairwise distinct $(n-1)$-dimensional subspaces of $\mathbb{R}^n$, then $\mathrm{bd}\, K$ is an ellipsoid.*

The following example from [1] shows that the condition on central symmetricity of K is essential in Theorem 12.5.3.

Example 12.5.4. For any scalar $-2 \leq \lambda \leq 2$, the set

$$K_\lambda = \{(\xi_1, \xi_2, \xi_3) : \xi_1^2 + \xi_2^2 + \xi_3^2 + \lambda \xi_1 \xi_2 \xi_3 \leq 1, \max\{|\xi_1|, |\xi_2|, |\xi_2|\} \leq 1\}$$

is a convex body in $\mathbb{R}^3$, which is elliptic with respect to any of the three planes given by $\xi_1 = 0$, $\xi_2 = 0$, and $\xi_3 = 0$, respectively. On the other hand, K_λ is not an ellipsoid if $\lambda \neq 0$.

The next result from [31] complements Theorem 12.5.3.

Theorem 12.5.5 ([31]). *If a convex body $K \subset \mathbb{R}^n$ is elliptic with respect to four pairwise distinct $(n-1)$-dimensional subspaces of $\mathbb{R}^n$, then $\mathrm{bd}\, K$ is an ellipsoid.*

References

1. Alonso, J., Martín, P.: Some characterizations of ellipsoids by sections. Discrete Comput. Geom. **31**, 643–654 (2004)
2. Alonso, J., Martín, P.: Convex bodies with sheafs of elliptic sections. J. Convex Anal. **13**, 169–175 (2006)
3. Alonso, J., Martín, P.: Convex bodies with sheafs of elliptic sections. II. J. Convex Anal. **14**, 1–11 (2007)
4. Auerbach, H., Mazur, S., Ulam, S.: Sur une propriété caractéristique de l'ellipsoïde. Monatsh. Math. **42**, 45–48 (1935)
5. Berger, M.: Geometry. II. Springer, Berlin (1987)
6. Bianchi, G., Gruber, P. M.: Characterization of ellipsoids. Arch. Math. (Basel) **49**, 344–350 (1987)
7. Blaschke, W.: Kreis und Kugel. Viet, Leipzig (1916)
8. Bonnesen, T., Fenchel, W.: Theorie der konvexen Körper. Springer, Berlin (1934).
9. Brunn, H.: Über Kurven ohne Wendepunkte. Habilitationschrift. Ackermann, München (1889)
10. Burton, G. R.: Sections of convex bodies. J. London Math. Soc. **12**, 331–336 (1976)
11. Busemann, H.: The Geometry of Geodesics. Academic, New York (1955)
12. Carathéodory, C.: Über den Variabilitätsbereich der Koeffizienten von Potenzreihen, die gegebene Werte nicht annehmen. Math. Ann. **64**, 95–115 (1907)
13. Gruber, P. M., Höbinger, J. H.: Kennzeichnungen von Ellipsoiden mit Anwendungen. In: Fuchssteiner, B., Kulisch, U., Laugwitz, D., Liedl, R. (eds.) Jahrbuch Überblicke Mathematik, pp. 9–29. Bibliographisches Institut, Mannheim (1976)
14. Heil, E., Martini, H.: Special convex bodies. In: Gruber, P. M., Wills, J. M. (eds.) Handbook of Convex Geometry, pp. 347–385. North-Holland, Amsterdam (1993)
15. Höbinger, J.: Über einen Satz von Aitchison, Petty und Rogers. Dissertation, Technische Hochschule Wien, Wien (1974)
16. Klee, V. L.: Some characterizations of convex polyhedra. Acta Math. **102**, 79–107 (1959).

17. Kubota, T.: On the theory of closed convex surface. Proc. London Math. Soc. **14**, 230–239 (1914)
18. Kubota, T.: Über die konvexe geschlossene Fläche. Sci. Rep. Tôhoku Univ. **3**, 277–287 (1914)
19. Kubota, T.: Einfache Beweise eines Satzes über die konvexe geschlossene Fläche. Sci. Rep. Tôhoku Univ. **3**, 235–255 (1914)
20. Montejano, L., Morales, E.: Variations of classic characterizations of ellipsoids and a short proof of the false centre theorem. Mathematika **54**, 35–40 (2007)
21. Nakagawa, S.: On some theorems regarding ellipsoids. Tôhoku Math. J. **8**, 11–13 (1915)
22. Nakajima, S.: Über konvexe Kurven und Flächen. Tohôku Math. J. **29**, 227–230 (1928)
23. Osgood, W. F., Graustein, W. C.: Plane and solid analytic geometry. Macmillan, New York (1942)
24. Penna, M. A., Patterson, R. R.: Projective Geometry and its Applications to Computer Graphics. Prentice-Hall, Englewood Cliffs, NJ (1986)
25. Petty, C. M.: Ellipsoids. In: Gruber, P. M., Wills, J. M. (eds.) Convexity and its Applications, pp. 264–276. Birkhäuser, Basel (1983)
26. Soltan, V.: Convex bodies with polyhedral midhypersurfaces. Arch. Math. **65**, 336–341 (1995)
27. Soltan, V.: Affine diameters of convex-bodies–a survey. Expo. Math. **23**, 47–63 (2005)
28. Soltan, V.: Convex solids with planar midsurfaces. Proc. Amer. Math. Soc. **136**, 1071–1081 (2008)
29. Soltan, V.: Convex solids with homothetic sections through given points. J. Convex Anal. **16**, 473–486 (2009)
30. Soltan, V.: Convex quadrics. Bul. Acad. Ştiinţe Repub. Moldova. Mat. **3**, 94–106 (2010)
31. Soltan, V.: Convex solids with hyperplanar midsurfaces for restricted families of chords. Bul. Acad. Ştiinţe Repub. Moldova. Mat. **2**, 23–40 (2011)
32. Tietze, H.: Über Konvexheit im kleinen und im grossen und über gewisse den Punkten einer Menge zugeordnete Dimensionszahlen. Math. Z. **28**, 697–707 (1928)

Index

B. Toni et al. (eds.), *Bridging Mathematics, Statistics, Engineering and Technology*,   147
Springer Proceedings in Mathematics & Statistics 24, DOI 10.1007/978-1-4614-4559-3,
© Springer Science+Business Media New York 2012